ENTROPY THEORY OF AGING SYSTEMS

Humans, Corporations and the Universe

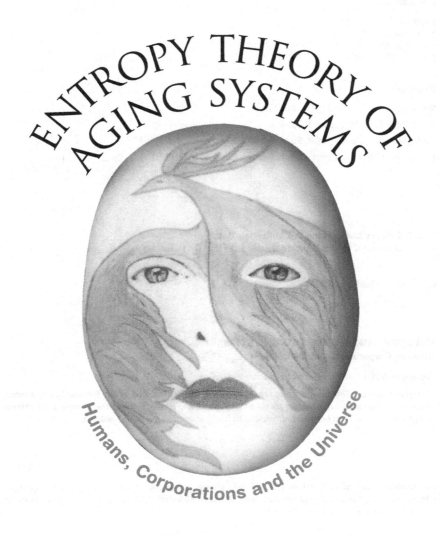

ENTROPY THEORY OF AGING SYSTEMS

Humans, Corporations and the Universe

Daniel Hershey

University of Cincinnati, USA

Imperial College Press

Published by

Imperial College Press
57 Shelton Street
Covent Garden
London WC2H 9HE

Distributed by

World Scientific Publishing Co. Pte. Ltd.
5 Toh Tuck Link, Singapore 596224
USA office: 27 Warren Street, Suite 401-402, Hackensack, NJ 07601
UK office: 57 Shelton Street, Covent Garden, London WC2H 9HE

British Library Cataloguing-in-Publication Data
A catalogue record for this book is available from the British Library.

Cover image: Barbara Hershey

ENTROPY THEORY OF AGING SYSTEMS
Humans, Corporations and the Universe

ISBN-13 978-1-84816-292-1
ISBN-10 1-84816-292-8

Typeset by Stallion Press
Email: enquiries@stallionpress.com

Printed in Singapore by Mainland Press Pte Ltd.

Dedication

Growing up on the block, on Elton Street, in the East New York section of Brooklyn, I knew I was different. What kid spends the summer sitting on a bench reading "Psychology of Sex", by Havelock Ellis?

At Cooper Union, as an undergraduate in chemical engineering, I discovered artists: students who seemed to be who I was, or wanted to be. Art and Engineering?

As a young professor at the University of Cincinnati I met my dream, Barbara, an undergraduate art student and photographer, a deep thinker. We married, merged, grew together, worked together, raised two lovely children together.

But she died suddenly, of leukemia, at the age of 48, at the top of her game, a professor at Miami University. I miss her so very much.

So here's to you, Barbara, the one love of my life, my inspiration, wherever you are.

Daniel Hershey
January 1, 2009

Contents

Part V. Entropy Theory of Aging Systems: The Corporation

Part VI. Entropy Theory at Aging Systems: The Universe

Introduction

In simple terms, entropy is a measure of order and disorder, in our human bodies, in so-called inanimate organizations such as corporations, and even the universe. If left alone, these aging systems go spontaneously from low entropy and order, to high entropy and disorder. From life to death, where death is maximum disorder or maximum entropy. This means we can also tell the direction of time (time's arrow) since natural things proceed in the direction of increasing entropy.

So we present here the commonality of principles which govern the birth, maturing, and senescence history of aging systems, all of us. We will show the entropic processes at work in humans, corporations, and the universe. We will show the entropy calculations which trace the lifecycle of everything. And in the end, it's all about life and death, Infinity and God.

In Part I, "Life and Death", Chapter I, we present two fictional short stories. The first, "A Journey", deals with death, what it feels like, and where we might go afterward. The second, "In God We Trust", tells us about George, the carbon atom, and his journeys through our world, and his commitment to the rationality of it all.

In Part II, "Entropy, Infinity and God", Chapters 2 to 4, we summarize the significance in our everyday lives of these three defining concepts.

In Part III, "Lifespan and Factors Affecting It: Humans", Chapters 5 and 6, we list the losses with age, theories and speculation in regard to the aging process, and the influence of increasing entropy, the metabolic rate, and exercise on human longevity.

In Part IV, "Entropy Theory of Aging Systems: Humans", Chapters 7 to 9, we explore an entropy theory of aging, including a new entropy age scale. We explore the influence of time in the life process, give more details on the confluence of the basal metabolic rate and entropy concepts in the aging and death of humans. All this leads to Excess Entropy and

Excess Entropy Production as bio-markers and predictors of life's longevity potential.

In Part V, "Entropy Theory of Aging Systems: The Corporation", Chapters 10 to 15, we introduce the aging of so-called inanimate systems such as cities, corporations and civilizations. We connect these aging systems to the entropy concept: why they age, the dynamics of the change process, and order and disorder. What we have is a new dialogue for understanding our existence. These so-called inanimate systems seemingly evince the same characteristics as living, aging systems. We analyze the corporate structure, how it functions, the efficiency of information flow, and how we use an entropy theory to characterize the aging process and its lifecycles.

In Part VI, "Entropy Theory of Aging Systems: The Universe", Chapters 16 to 18, we develop entropy concepts and equations governing the Universe. The birth, maturing, and death of the Universe can be understood and traced, from an entropy point of view. With pressure, volume, temperature, and entropy calculations, we can understand the lifecycle of the Universe. And its origin and return to Infinity.

Thanks to graduate students Basak Bengu, Elif Bengu, Pablo Rosales, and Biao Qi at the University of Cincinnati, for assisting in the organization and editing of the manuscript.

PART I

Life and Death

CHAPTER 1

Two Short Stories

A Journey

I died last night and it was very interesting. Not nearly as bad as I thought it would be; quite a surprising experience you might say. It takes a moment of pain, pretty much like a pin prick, after the last breath, for our bodies to turn off all those intricately connected switches, sending a signal from the brain to all parts, that this is the end. First the brain dampens its buzzing circuits, then the heart ceases its contractions much as an engine grinds to a halt after the power is shut off and finally muscles, blood vessels, skin and bones relax. The word gets out to our red and white cells. The foreign bacteria and viruses who exist parasitically on and within us get the message. It is over; the system is inoperable. It is time to get out.

And I got out too. That's what I meant by this death experience being a surprising event. I never dreamed that there was more to me than skin and bones. But when I died I too departed, simply and easily, unshackled from the old carcass just as smoothly as slipping and sliding down a greased chute. You drift out, much as air leaves a slow leak in a tire. And before you know it, there you are, ethereal and invisible, floating around as a packet of energy and essence. Oh you know who you are, and who you were but it doesn't seem to matter. You are disengaged from the anchor and free. And it dawns on you that it was only a temporary thing, this occupation of the cumbersome, burdensome, corporal being.

I noticed one important fact right away. I was lighter than air and in no time was ascending up above the trees, past the tallest canyons of buildings, into clouds that now seemed abrasive and richly textured. But it was a simple matter for me to sway and bend with the tides of turbulent winds, all the while steadily penetrating upward until I broke through the clouds into the warm and refreshingly clear upper atmosphere and beyond. Much as a

3

balloon will float to an altitude which balances its buoyancy forces, so too I found my level in a region far from earth. And there to my surprise I found kindred souls, all of us units of energy and consciousness, identifiable by these characteristics much as before, when we knew each other by height, weight, skin color and voice. Depending on your energy density, you might settle in one region or another, nearer to or further from the earth and hence closer to or farther from the sun.

We live here for an indefinite period as far as I can tell. We'll stay until our energy levels are depleted, I'm told. And then? Well no one seems to know what comes next except there is a rumor that after the fun and relaxation of this new life, we are to be assigned another mission on earth (by God who is in charge) though no one can remember any previous transfers. It seems here we energy packets have no memories.

In God We Trust

(I) George was just an ordinary carbon atom, not too heavy, not too tall. His weight was exactly 12.01, the same as most of his fellow carbons. His spin orbitals were properly arranged, with two electrons in his inner K shell and four unpaired electrons in the L shell. His nucleus contained six neutrons and six protons as required, all nicely packed. Sure he knew some carbon atoms who were a little heavier, but they only constituted about one percent of all the carbons. George found contentment in knowing that he and his kind were very constant.

And they were everywhere, these carbon atoms, among the other hundred or more kinds of atoms in the world. Unlike the noble atoms like helium, neon and argon who disdained to join with the other atoms, carbons could be either alone and free and unattached or combined and married to other atoms to form family units. Alone, or joined to other carbon atoms, they could form the very durable and organized bonds of diamond or the less ordered, softer structure of graphite. Everyone said that the married carbon atoms were the fulfilled ones, for in marriage they transcended themselves and became part of an almost infinite variety of family units, each distinctive in size and weight and activity. Some, like the carbon-oxygen family were light and adventurous and floated beyond their substrate. Others when combined with metal atoms and other units became heavy and large and conservative and hardly capable of moving from one neighborhood to another.

It was a simple life which George (the ordinary carbon atom) led; mostly he was concerned with maintaining a stable, resting state for his electrons, trying to control his energy and avoiding those unseen forces which could destroy him. But this wasn't easy, for in the world there were mysterious forces at work which could disrupt his life. In ways which he and his fellow carbon atoms did not comprehend and at unpredictable times there would be unleashed from the heavens, storms of energy, sometimes as electrical flashes, and sometimes as great vibrations. Very destructive and dangerous were the high speed helium atoms, which could rip you apart. So could high speed electrons. And there was gamma radiation, the electromagnetic waves. In waves and quanta these pulses of energy would come, blanketing the world as if some holocaust was about to descend. But not everyone was affected. (Certainly the imperious, noble helium, argon and neon would be immune.) While most of the atoms shuddered during these storms, the unlucky ones were mesmerized and then activated. Their electrons jumped, their nuclei quivered as if in transfiguration, and in a frenzy they mated and formed family units. Now separated from their past, these new compounds no longer resembled their old cohorts. Sometimes the family units, in some preordained ritual, would join forces in a communal existence, sharing their new bonds.

George was a religious atom, as were most of the atoms he knew. He believed in an organized universe with his God in control. He accepted his life as an atom philosophically and was willing to be mated when his time came. He had blissful confidence that God was good, and that there was a hereafter. If God chose George to be mated, he would go willingly. He knew that ultimately God would return him to his single state. So George was a believer and found the inner strength to endure. Those terrible storms which arose from time to time were acts of God; God has a purpose.

On this warm and sunny summer day, at this time in history, George a carbon atom in amorphous association with his colleagues in a charcoal briquette, was thrust upon a metal grid, doused with a volatile liquid, exposed to a lit match and engulfed in searing flames.

"My God, why have you chosen me for your torture?" cried George. His electrons were excited; the stress of heat and temperature produced an instability that suggested the end of the world. And when it seemed that he would not survive, two oxygen atoms, also excited and vibrating sympathetically were suddenly there, with their electrons bound to his. And they were mated to produce carbon dioxide.

To be a carbon dioxide molecule is to be a colorless gas, representing a small elite group on earth. For George it meant a loss of identity, a subordination of his personality for the good of his family, and a sharing of his electrons which were so precious to him. But despite these compromises, and even allowing for the loss of his carbon cohorts, there was a positive side to his marriage; he was no longer a one-dimensional character, no longer immobilized. He could soar with the wind, dive into liquids and dissolve, diffuse through membranes and do things he never dreamed were possible. "God has been good to me; I am seeing and experiencing miracles," he thought. "The God that brought the fire which caused me to suffer so much must have been testing me. He found me worthy, for have I not led a devout life. Now he has brought beautiful oxygens to my side. Surely God is good."

But the freedom which George and his oxygens felt, the euphoric flights were to be short-lived. Unnoticed near the flames were immense mountains which alternately expanded and contracted, inhaling and exhaling huge quantities of air in the process and generating terrible wind storms. It was during one of these cycles that George was sucked into the cavernous interior of one of the mountains, pressed against the moist, sticky walls by the vacuum and forced through the membrane which was the outer layer of the walls and dumped into one of the cells.

Then the attack by exotic molecules began. First it was a pyruvic acid molecule, which surrounded George and his oxygens. Aided and abetted by catalysts which drained George of his resistive energy, the pyruvics easily weakened the bonds between George and his oxygens and dissolved the marriage. In the ensuing confusion there were some rapid rearrangements and when the struggle was over, George and his oxygens were tied to a pyruvic acid molecule. It was just the beginning of the nightmare for George, who now was being whirled rapidly and unstoppably in the cell. In this roller-coaster ride in the cell he mated temporarily with hydrogen and water, lost electrons and got them back, and was subjected to deforming physical stresses — until, dizzy and exhausted, George regained his senses to find himself part of an almost infinite sea of carbons, oxygens, nitrogens, phosphoruses and hydrogens. He was now but one miniscule part of a DNA molecule, one carbon atom bound to thousands of others in chains and helices, in a chemical prison. "Being part of DNA is to be the controller of the life process," George was told. "We make the food and building blocks and supply the genes who enable our cell to survive and reproduce. Every atom is critical."

So he was welcomed to the team, with a warning, "No independence is allowed here; all must work for the common good. Get out of line or neglect the instructions being passed on by the messengers, and the whole process is upset. No mutants are tolerated in this cell, so there can be no laxity. We are all comrades."

(II) To be a virus is to be someone who cannot reproduce with your own kind. To be a virus is to be a parasite, living in host cells which supply your food for life and the means for reproduction. To be a virus is to be about one-twentieth of the size of the smallest organism capable of independent reproduction. It is not an easy life. You must live by your wits, but a virus is immortal as long as one of its species survives. Let one enter a host cell and identical viruses will be produced. Though new, they are the same as the old.

Mary accepted her life as a smallpox virus. She was larger than most and heavier. With a brick-like shape which gave her strength and a hexagonal protein head containing DNA (which was the reason for her intelligence), Mary could compete easily for the host cells against the other viruses who had poor quality RNA in their heads and weaker shapes.

She was born in a lung cell and quickly understood the peculiar existence of her kind — not being able to eat or engage in sexual activities in the usual way. To be with your fellow smallpox viruses was to look at your own image in a mirror. Why viruses were destined for this kind of life Mary attributed to the design of God.

"The God that invented the virus also created bacteria and the cells, so the viruses could survive," she often said. "The God which causes the flames and heat which kill viruses also gives them special shapes for survival. And when God brings a plague of antibodies to destroy us in the blood streams, there must be a reason. He intends us no permanent harm, for he also causes the virus to become immune to these same antibodies, guaranteeing our survival. Though there will be times when living conditions become intolerable, God will always provide the means for survival; viruses are the chosen ones." Mary believed in God, in His wisdom and omnipotence.

Mary the smallpox virus, born in a lung cell with hundreds of her sisters, in response to some primordial pressure, broke out of the cell of her birth and settled onto the exposed surface of the living lung. There she rested but briefly, for suddenly the ground wrenched, a distant rumble became audible and the wind velocity increased. Hugging the surface with her powerful legs, Mary cringed in fear of this irrational event, a storm which threatened her life. In a fraction of a second a perfectly calm day produced a cataclysmic burst of wind which swept her out of her home and into the air. She had been expelled from the land of her birth by forces governed by her God, for reasons unknown to her.

Sailing through the air, seemingly weightless, Mary had time to observe this new world. In some frightening way she was exhilarated by the danger and freedom, neither cancelling the other exactly. "Has God a special mission for me?" she wondered as she floated by a charcoal grill which was aflame. The heat was intense but she was beyond its reach. "I don't have much time to find a home. Where will I land?"

Mountain ranges loomed before her, positioned uniformly around the flames. Mary marveled at the symmetry of the landscape, thoroughly enchanted with the new vistas and unaware that she was rapidly approaching one of the mountains. She didn't notice the periodic updraft which was drawing her toward the mountain top. The suction was getting stronger, the turbulent wind velocities were accelerating, her speed was increasing and she was now being buffeted badly. The joy of floating in this new world rapidly gave way to despair and an ineffective attempt to stabilize her flight. Sensing the hopelessness of her struggle, aware of the God-like dimensions of the forces at work, she succumbed to their will.

Mary did not crash into the mountain, but was swept instead into one of its cavities, impelled by a vacuum which also drew air, other viruses and bacteria. She crashed onto a lung surface and instinctively burrowed into its interstices. "I'll be safe from the wind here, but my strength won't hold up much longer without a host cell," she thought. "I seem to have done this before; all this seems so familiar." It was as if she were programmed by some grand designer.

Looking only for something familiar and comfortable she finally contacted a cell — a lung cell — and was surprised when the physical contact kindled a sexual stimulation previously unknown to her. Through her pores Mary exuded an enzyme which weakened the host cell's membrane and rendered it more permeable. With her legs firmly attached to the cell she squatted, contracting the external sheath of her body and allowing her inner tube to penetrate the cell membrane. Now straining mightily, with a killing effort, Mary drained the DNA genetic material from her head, through her body tube and into the host cell. The process took only one minute but now Mary understood that this was her reason for existing. Through her the smallpox virus would continued to live: God had chosen her for the most important function of the universe. She also knew that by her act she would die, for stripped of her DNA she was nothing. But death meant others would be born soon in her image. From Mary would come hundreds of smallpox viruses and she would be reborn. Hallelujah! She was immortal!

Old Mary died, but her DNA seeds were firmly implanted inside the host cell. Within minutes, the carbon atoms, the oxygens, nitrogens, phosphoruses and hydrogens of her DNA essence spread into the genes of the host cell. In less than one hour they had found their counterparts and had taken over the operation of the cell. Where formerly the DNA of the host cell had peacefully manufactured replacement parts, now they were overcome and the process was shut off. Instead of a pastoral existence, they were bullied into producing replicas of the smallpox virus, under exact instructions from Mary's messengers. First the DNA was to be assembled and condensed into a head. Next the legs and body were to be made and finally everything had to be covered with protein coat. It took a few hours to complete the operation and now hundreds of new Marys were contained within the spent host cell. Old Mary was dead, young Mary lived.

"A carbon atom is passive by nature," said George to a fellow atom as they worked on the DNA assembly line. "We're usually un-reactive unless highly stimulated. I accepted my fate when God converted me to carbon dioxide, and I can see a purpose in His drafting me into the DNA of this cell. The work is dull here but at least I feel part of a team. And we seem to be doing useful work. But when those dreadful smallpox atoms come in and force us to do their work, that's going too far. I will not build Mary viruses; these smallpox viruses are evil monsters who live parasitic lives, draining the life out of docile cells in their effort to proliferate. These fascists are also trying to force me to join one of their DNA molecules and become part of them. I'll not do it!"

But George didn't have much choice in the matter. He was forced to do the bidding of the smallpox messengers, assembling their DNA, until about halfway through one production run when the orders came for George to join the DNA which was being assembled as the head of a Mary virus. There is no resistance to such instructions since a virus enzyme is always present to break the bonds to recalcitrant atoms. So George the carbon atom was integrated into a smallpox virus.

"God is surely testing me in this new role he has chosen for me," said George, unhappily. "These are Godless creatures, and for me to aid in their destructive activities is offensive. I am a prisoner for reasons which God only knows."

Young Mary was beginning to stir, sensing that she was now complete as a smallpox virus.

"Thank you, God, for continuing the miracle of creation which proves that you are happy with us," she said. "I hope I may be worthy of your trust and can carry out your wishes. I pledge my DNA to your service."

George and Mary, Mary and George, were inseparable. George in slave labor, working against his will in the smallpox DNA helix, did his job as carbon atoms must. No carbon atom is unstable; carbon atoms do not break their bonds easily. He prayed that God would rescue him and restore his freedom. Young Mary, full of early bloom, energetic and fearless, probed at the host cell's surrounding membrane in an effort to escape. She found a weakness in the membrane of this dying cell and easily plunged through, ready for adventure.

"God give me the strength and courage to persevere," she said as she emerged from the cell.

(III) Mark had enjoyed sitting around the charcoal grill with his fellow students, eating the hamburgers and chicken. All of them were new to the university so it was fun to swap horror stories about registration difficulties and new professors. They represented the U.N. in microcosm, having come from so many different countries. One of the animated discussions they had that day was about religion, and whose was best. The Buddhists, Muslims, Hindus, Christians and Jews made their representations and when it was over, they agreed to disagree, each strongly determined that his religion was best (and contained the only truth about God).

Ever since that picnic a week ago, Mark felt a bit feverish, and now he noticed that chills, shaky spells and restlessness were added to his symptoms. He awoke one day with a severe backache, vomited and found he had a temperature of 105 degrees. There were faint, irregular blotches on his skin. His doctor told him the bad news: he had contracted smallpox; apparently his lungs were the primary infection site.

"What rotten luck," Mark said to the doctor, knowing he would now have to miss some critical classes at the university. "How the hell did I get it?"

"Probably through one of the fellows from India or Africa or South America who were with you the day of your picnic."

"God is punishing me for not defending him more forcefully," Mark said sardonically.

The smallpox infection proved to be more virulent than they suspected; Mark had hemorrhagic smallpox which specifically affects the lungs and is the most dangerous form of the disease. In addition to the general symptoms and the skin lesions, and the damage to the liver and spleen, there is bleeding in the lungs with pneumonia arising as a secondary infection. Mark had never been in good health; as a student away from his parents he had further weakened himself by eating poorly. Thus this attack of smallpox became a very serious matter. Despite the best of medical care, despite inoculations to counteract the smallpox virus, Mark gradually slipped into a coma, the virus now too entrenched to be overcome by Mark's antibodies. He died two weeks after the initial infection.

"Why has God punished us?" asked his parents. "Mark was so young, so full of promise. His whole life was before him; there was so much he wanted to do with his life. What kind of God is this?"

"Isn't it ironic," said the mourners, "how it is always the good people whom God chooses to die early."

"Our duty is not to question God's will," replied the others. "He has a purpose, even if it is not obvious to us."

(No one asked whose God had killed Mark — whether it was the God of Buddha, or Jesus or Abraham. Or was it Mary's God or George's? So they laid Mark into the ground, hoping for a sign that this death was in God's name and would be for the betterment of mankind.)

(IV) Oh, how merry they were. A perfect climate for Mary viruses, they moved into one cell after another of Mark's lungs, organizing this cell and that one, generating more and more smallpox viruses at an accelerating rate. (George, on the DNA assembly line, was forced to work at a maddening pace, hardly daring to rest or think; there was only time to follow the orders of the messengers.) Everyone was so busy and full of robust health during those two weeks they didn't notice that the cells they invaded possessed increasingly sluggish DNA. Soon George and the others were having a difficult time converting the host cell's DNA for the manufacture of Mary viruses.

And then the alarm went out, "The new cells are not operable! The new cells can no longer be organized to produce smallpox."

And then the panic developed, "We'll starve unless we find more cells!"

They foraged in the liver and spleen, traveled the entire arterial and venus routes and even tried the lymphatics. It wasn't easy, moving through blood and lymph fluids, for then they were vulnerable to attacks by antibodies. But their ranks were swollen by the urgency of their predicament and Mark's antibodies were unable to eliminate all the Mary viruses.

Two weeks after Mark was invaded by Mary he died rendering all of Mary's host cells useless. The bacterial flora from Mark's intestines now migrated to the lymphatics, blood capillaries and veins and finally into body tissue including the respiratory system. The aerobic bacteria used up the available oxygen, allowing the anaerobic bacteria coming mostly from the intestines to begin their proliferation. New forces were now at work; strange enzymes and chemicals appeared which were obnoxious to Mary. The new bacteria which appeared were immune to Mary's enzymes. The world was collapsing; they were on the verge of disaster. Chemical bonds which were unbreakable before, now crumbled before the new enzymes. With no host cells and a toxic environment, the smallpox viruses began to die in hordes. Some even became cannibalistic and attacked each other in their death frenzy.

"The God who was so good has turned on us," said Mary. "This ugliness He has given us must be punishment for some offense we have committed against Him. Surely there must be a reason for this cruelty."

The Mary viruses died seeking some sign from God. "Thank God the smallpox viruses have been destroyed," said George. "I had my doubts about Him, but I guess we atoms cannot comprehend the infinity of God's deeds."

The new enzymes broke George's DNA bonds and liberated him. He eventually mated with four hydrogens to become an odorless, colorless, gaseous methane molecule and floated away from the crumbling smallpox milieu, up through the earth of Mark's grave into the free air.

"Once again a gas? Isn't this a bit redundant?" George thought. The sensuality of his new marriage to the hydrogens was dampened by the memory of his previous experience. George was thankful for his release from the Mary virus but remained leery.

"I'll do God's bidding, but I hope He chooses someone else."

George's methane family, driven by the winds, drifted about the earth, settling occasionally in a coal mine or a pool of stagnant water or some sewage. (Within the time frame of the universe it was a relatively short time.) And finally, at a time unmarked, in a place unknown, lightning flashed at George. The searing heat tore at George's methane bonds and in the presence of a few oxygens, caused a chemical reaction to form free carbon and two water molecules. The odyssey was ended, George was again a free carbon atom.

"Thank God."

(V) Alice loved to play with her Milky Way, her pinwheel-shaped galaxy with its five spiral arms, a flattened disc of two billions stars, gases and cosmic dust. She delighted in displacing the constellation Sagittarius from its center position, a movement which produced resonances within the Milky Way and upset the fine balance of magnetic and electrical forces. This in turn would alter the distance of Sagittarius from the earth's sun, causing sun spots which affected tiny earth so interestingly.

God has been good to Alice.

PART II

Entropy, Infinity and God

CHAPTER 2

Entropy

A quick summary of what we mean by entropy:

Everything you wanted to know but were afraid to ask

- We are born organized, age with increasing disorder, and die in maximum disorder.
- We are born with low entropy, mature with increasing entropy, and die at maximum entropy.
- The driving force for life, then, is the entropy distance from maximum entropy.
- Death is the ultimate disorder, maximum entropy.
- Entropy can tell us something about time's arrow, about the direction of time.
- Entropy maps the degree of order and disorder: higher entropy indicates more disorder.
- Entropy is a measure of the disorder in an organization.
- Entropy increases as differences or tensions within the organization are dissipated.
- Entropy increases as the number of organizational units is increased.
- Entropy tends towards a maximum in the vicinity of death, as control is lost.
- The corporation tends toward maximum entropy when all units are independent and equal (a disaster or chaos condition).
- Entropy decreases to a minimum, ideal value, when all corporate units have perfect access to the leader and to each other.
- Tendencies toward verticality in organizational structures increase entropy and drive efficiency down.
- Excess Entropy (EE) is a measure of the entropy distance from disaster: your present entropy condition minus maximum entropy.

- EE achieves its greatest value and furthest distance from disaster when an organization moves towards an ideal structure.
- Excess Entropy Production (EEP) is the rate of change with time of Excess Entropy (EE). EEP diminishes with age and nears zero in the vicinity of death. For the aging corporation, a diminishing EEP can signify stagnation, and a general decline in organizational vitality.
- Achieving maximum entropy is a randomizing process. It means landing in the most probable state, the universal attractor.
- Entropy characterizes symmetry. Higher entropy means more randomization.
- The orderly decay, the aging process, is entropy driven, towards maximum entropy.
- A gain in entropy means a loss of information.
- Entropy describes disorder and our ignorance.
- Maximum entropy is when everything is equal, when there is a total randomness.

Higher Entropy	Lower Entropy
Random	Non-random
Disorganized	Organized
Disordered	Ordered
Configurational Variety	Restricted Arrangements
Freedom of Choice	Constraint
Uncertainty	Reliability
Higher Error Probability	Fidelity
Potential Information	Stored Information

- Organizations, such as corporations, demonstrate entropy, structure, and information flow.
- Structure is a requirement for the storage of information, for without structure all the information available is simply potential information, theoretically available only when the system organizes itself.
- We can calculate the entropy content of a corporate structure and conclude a certain lifecycle and behavior.
- If all the units in the organization overlap in their duties (do the same things), this leads to high entropy and much informational disorder.
- If the structure is very vertical, we have an inefficient structure in terms of information flow, since knowledge at the bottom must pass through many units in ascending towards the leader. In passing through so many

units, the information naturally becomes distorted, increasing in entropy. A more horizontal structure would be better.

- If information doesn't flow easily through the layers of direct responsibility, bypassing the bosses leads to a shadow organization, which allows information to short-circuit the boss.
- Decision-making is easy and organizational efficiency is highest, in the short run, when most of the power resides near the top of the structure.
- Allowing major power to be concentrated in the lower rungs of an organization diminishes the efficacy of the structure, introduces chaos and information distortion and entropy increases.
- Allowing all the units in an organization to have equal authority, to be equal in all aspects, and to be completely independent is a clear prescription for disaster and maximum entropy.
- Distributing the power equally among the units, also raises entropy. We need a certain tension of life, to produce order and structure and low entropy.
- If all units are equal in power and activities, and are totally independent, we have not only a prescription for disaster, but also a definition of organizational equilibrium and maximum entropy and organizational death.
- We can store information and control with structure at low entropy, or we can evolve towards only potential information and no structure, at high, maximum entropy and death. And of course, size, (bigness), in the corporate world comes with the baggage of high entropy.
- Living organisms and the so-called inanimate systems such as corporations, countries, and civilizations are similar dissipative organizations, living, growing, evolving, aging, and dying as open systems, entropy producers, striving mightily to stay away from maximum entropy, going from one stationary state (at minimum entropy production) to the next stationary state (at another minimum entropy production), to the next, and the next, etc.

CHAPTER 3

Infinity

To understand entropy we need to consider its relationship to Infinity.

To understand the aging of the universe we need to "feel" Infinity.

Infinity is everything, a place, a condition, a destination, a location of infinite information.

- I can take an ordinary book, with its letters organized into words which are strung in sentences on the pages which are bound together in a book, whose volume is contained in a library of books. I can rip these pages from the book, and with my scissors cut each page into individual words and each word into individual letters piled on the floor. And what have I? If you think about it, you'll realize that I have more than just a pile of junk letters, the remnants of the organized assembly of letters which became the words and the sentences for the plot of the story of the book. For what that pile of individual letters contains is a near Infinity of information, the potential for writing not only the original book, but many, many other books. Simply by rearranging the letters piled on the floor. That pile of letters is like the infinite potential information of Infinity, from which is obtained the organized and stored information of our book. The book and the library represent the finite world of our limited intelligence.
- Infinity contains an infinite amount of potential information.
- Infinity contains all the information we know, or can ever know, or can never know.
- I'll count my numbers. Count with me: 1, 2, 3, 4, ... Give me your largest number. Is it a billion? One zillion? Or if you work with numbers and play mathematical games, is it $10^1, 10^2, 10^3, 10^4, \ldots$? Stretch for the largest number you can imagine, and I can add one to it. One zillion, you say, and I respond one zillion and one. One zillion zillion, you say, and I

counter with one zillion zillion and one. And where does this game end? In Infinity, of course. But where is Infinity?

- Sit quietly and contemplate Infinity. Picture in your mind an expanse, up to the horizon. Now, in your mind, remove the horizon. It feels like nothingness without the horizon, without some frame of reference. Picture in your mind the sky, up to the stars. Now, in your mind, remove the stars, and the moon, and the sun. It feels like nothing without them, without some frame of reference.

- Being in Infinity is to be without all frames of reference, including the horizon, the stars, and all other tangible signposts. Being in Infinity is being without limits. It is just being, without corporeal frames of reference, without temporal frames of reference, without physical and mental boundaries.

- To be in the infinite state, clear your mind — clear your mind's eye — of all tangibles. You'll be left with emptiness, open space stretching beyond everywhere, expanding beyond everything. Infinity may at first seem empty, but it isn't, for it contains all essences and information imagined and not yet imagined.

- It is useful to ask, how did it all begin, but then we realize Infinity didn't begin. Asking the question demonstrates the limitations of our human brains, which require a beginning, a middle, and an end.

- It is useless to ask, what is it, because Infinity isn't related to anything we can know. Infinity simply is.

- It is useless to ask, what is beyond Infinity, because there is nothing beyond Infinity.

- It is useless to ask, what are the coordinates of Infinity, because Infinity is beyond coordinates.

- It is useless to ask, whether God controls Infinity, because in actually, Infinity transcends God.

- We have great difficulty "seeing" Infinity. Mostly it has to do with the limitation of our human brain. Mostly we exist through our finite senses of touch, sight, smell, taste, and hearing. Something exists for us if it has dimensions, and heft, and significance. Even an intangible concept, or an emotion, can be defined in terms of its impact on our senses. Our thoughts are circumscribed by our experiences, allowing us to corral these esoterica within a fence, or wall, or membrane. We need to do this in order to understand — or try to understand — our existence. We need to make sense of life and death.

- Infinity has more to do with a sense of "being" and that's hard to deal with. Infinity just "is".
- To understand Infinity is to deny ends and endings.
- To understand Infinity is to deny boundaries.
- To understand Infinity is to deny beginning and beginnings.
- To understand Infinity is to deny place and places.
- To understand Infinity is to deny focus and focusing.
- To understand Infinity is to deny existence and existing.
- To understand Infinity is to deny limits and limitations.
- To understand Infinity is to allow softness and vagueness.
- To understand Infinity is to allow expansion beyond limits.
- To understand Infinity is to allow that God represents a synonymy.
- To understand Infinity is to allow that all concepts of God are valid.
- To understand Infinity, don't try to imagine what it looks like.
- To understand Infinity, don't try to imagine where it is.
- To understand Infinity, begin with an ether which fills all voids.
- To understand Infinity, focus on a sense of being.
- Infinity must be a universal something, and a universal someplace, and a universal being.
- Infinity is everything we know, or can ever know, or can never know.
- Beyond what we imagine, beyond what we can imagine, beyond what we can never imagine.
- In Infinity, a random accumulation of information at low entropy allows the condensation of vacuoles of matter and energy. And in forming, these vacuoles temporarily separate themselves from the substrate of Infinity. We condense from Infinity in the same way that steam condenses to water droplets and then to ice crystals.
- It was such a precipitated mass of information and energy that became our universe, began our lifecycle with a big "bang". Our universe, originating as information in a state of low entropy, has begun to expand and age, generating its own ambience and lifecycles. We were born in a low entropy state and have evolved through succeeding states with ever increasing entropy.

CHAPTER 4

God

If we try to understand entropy and Infinity and the origin and aging of the universe, God may be lurking nearby.

- What does God look like?
- What does a belief in God mean?
- Is God knowable?
- How could we look for (and find) God?
- What does death look like? Like Infinity? Like God?
- Is God male or female?
- Our invention of God, in whatever culture, in whatever religion, in whatever form, in whatever point of history, clearly represents our understanding of our limitations. God, the infinite. God, the all knowing. God, the good. God, the angry. God, the merciful. God, the doer. God, the passive. God, the distant one. God, who speaks. God, who remains silent. God, of love. God, of the sea. God, of plenty. God, the sun. God, the parent. God, the mother. God, the father. What we know, what we hope to know, what we can never know, God knows. What we feel, God feels. What we need, God provides. God fights wars, makes us fertile, allows us to die. God brings us home after death. God, the eternal, shine your light on us.
- We believe in God because we believe in Infinity. There is comfort in believing that God fills the gaps in our universe. There is discomfort in wondering what is beyond our universe. There is comfort in believing our hard times on earth will become a loving, warm afterlife. There is discomfort in contemplating a hellish afterlife if we do not measure up here on earth. There is comfort in having rules and structures controlling our lives. There is discomfort in encountering others who live by different sets of roles and structures. There is comfort in being a chosen people. There is discomfort in meeting others who are also the chosen ones.

25

- "Heaven and hell is within each of us." "I think, therefore I am." (On God, by God)
- "God is the principle law, the sum of all the external laws in existence. God is a material being, identical with and equivalent to the order which governs the universe." (Spinoza)
- "To know that what is impenetrable to us really exists, manifesting itself to us as the highest wisdom and the most radiant beauty, which our dull faculties can comprehend only in their most primitive forms — this knowledge, this feeling is at the center of all truly religiousness. In this sense, and in this sense only, I belong to the ranks of devoutly religious men." (Einstein)
- "My God is similar to the God of the mystics, a psychological truth experienced by each individual." (Jung)
- "If God does not exist, then all is permitted." (Dostoyevsky)
- "I am what I am." (God to Moses at the burning bush)
- "There is a God-shaped hole in the human consciousness, where God has always been." (Jean-Paul Satre)
- "Is this concept of God necessary to account for human destiny? A personal God who gives a heaven and hell? The belief in a spiritual being, with a propitiation of powers, and conscious seems to deny the validity of the Buddhistic profession that there is no supernatural being. A sense of religiousness gives us a feeling of sacredness, a fusion of inner and outer selves. Is God responsible for Luther, St. Francis, Jesus, Buddha, and Archimedes, Faraday, Mendel, Darwin, and Beethoven, Wren, Wordsworth, Van Gogh? God makes eternity available and makes it easier to endure the hardships on earth. God gives us birth, marriage, death, reproduction, suffering, comradeship, physical and moral growth, knowledge of sin, righteousness, absolution, communion, truth, virtue, beauty." (Huxley)
- We commune with God by self-organization. We construct temples and beliefs to be in harmony with God. Eternal God, we appeal to your absolute authority. You sanctify our marriages. God, our father, God the holy ghost, God, the son, God, the holy spirit. God, the sun, the Messiah who leads the chosen people. Who also leads Islam and Buddhism. God you give us unity, uniformity, continuity. You have the power over material things, you explain our world and our lives within it. You illuminate and direct our thoughts, feelings, wills, our minds. You give us life, and connect with an infinitude of things. God, you are infinite. God you are Infinity.

- We found you, God, and understand better our universe. God, you are present in all cultures. You bridge religion and science, from the natural to the supernatural, from the Egyptian priests to the powerful land owner. You've provided an intellectual definition in the Scriptures, in mythology. You are in many forms — river God, and dyads, such as Venus, Minerva, Mars.

- God, you give us a glimpse of the infinite. Or are you really Infinity? You transcend the natural. You are supernatural. You are so superior. God, the void. God, the enemy. God, the companion. You explain our world and our place in it. Freud, the skeptic, thought you were a phantom of an infantile father complex. Job, the questioner, elicited the quote from God, "I am the Lord, my ways are not your ways, you cannot understand the divine purpose." Ultimate reality is and always will be a mystery, to be feared as well as loved.

- There are many religions and beliefs in God, but there are also some absolutes: truth; beauty; goodness; holiness; unity. Whatever we profess, we are universally concerned with our destiny: our position and role in the universe, and how to maintain that position and fulfill our role. So we develop organizations — we self-organized — as a means of coping with this, through ideas, emotions, attitudes. Do animals have Gods? Are we them? Do we feel God? Wordsworth said, " I have felt/a presence that disturbs me within the joy/of elevated thoughts: a sense sublime/of something far more deeply."

- God is not immutable. God can be very silent and not speak (Book of Silence: Job, Lamentations, Ecclesiastes, Ester). Why did God create earth, then destroy it? God is baffling, irritating, inconsistent, arbitrary. Why does God have no father and mother? Sometimes God is not silent and talks to himself: "Let there be light… Let us make man in our image." God is angry with us and causes a flood, then says, "I will not cause the flood again." God, the destroyer. Muslims believe there is heaven and hell within each of us. Where did God come from? Did the Gods emerge, two by two, from watery wastes?

- God tells Abraham to be fertile, though he is at least 100 years of age. God befriends families: Jacob; Laban; Joseph. God is God-like. God is here and there, in the Exodus, the second book of the Bible, but is absent in the Book of Genesis. Moses felt God's exhilaration. God the liberator sent Moses out of Egypt. God, the lawgiver, defined ethics with the ten commandments. Does God age? Is God young and old? Can God die? Does God ever fail? Does God find God strange? Does God love? Are the first five books of the Bible true descriptions of God? Does God have a wife or a significant other? Or a husband? Does God eat, get hungry?

- Some say the beginning of wisdom is the fear of the Lord. Bertrand Russell calls God a fiend, for if life's outcomes are deliberate, then the purpose must be that of a fiend. God, the bystander (Book of Ruth). God, the romantic (Songs of Songs). God, the puzzle master: "Who can straighten what he has twisted (Ecclesiastes). Does God lose interest in us? Is God safely incarnated in the law (the mind of God), incarnated in leadership (the action of God), incarnated in prayer (the voice of God). Job, by being God's most perfect image, nearly destroys God. How could this happen?
- In the infinite, the concepts of Infinity and God merge.
- In Buddhism, no God descends to earth as in the monotheistic Christian and Muslim faiths.
- The God-image is a nonequilibrium force which causes evolution.
- Self-transcendence: we rise beyond our mortal, corporeal being into God's realm, into Infinity.
- God the creator, Infinity the essence.
- God in the mind, Infinity beyond the mind.
- God is the universe, Infinity is beyond the universe.
- God is excluded from science, Infinity is science and God.
- God gives us personal ethics, Infinity is universal ethics.
- God gives meaning to life, Infinity is life and beyond.
- God defines our existence, our death and rebirth. Infinity contains existence, death, and rebirth.
- Discontinuities are things that do not follow smoothly from before, these black holes of essence, thought, energy, mass, information.
- The concept of God is a singularity, a discontinuity.
- God upsets the laws of nature by interventions and miracles.
- Chaos may be a state approaching maximum entropy if there is no pattern discernable. And perhaps we've connected chaos, and our seemingly chaotic world, with a need to believe in God, for God is a sign, a symbol, of order and control. God gives us a deterministic chaos, in which the patterns and behaviors of our world are known only to God.
- Albert Einstein said, "God does not throw dice," which means Einstein believed that God delivered order, but it was our responsibility to discover this order.
- We ponder the immense questions of God, whose God, how many Gods, and the meaning of life and death, heaven and hell. If the God or Gods

of our universe exist, then they too must have evolved from Infinity. Perhaps our Gods were born when our universe was formed.

- The inhabitants of the newly formed universe know only that stored information which was inculcated into it at the beginning of a universe's existence. The occupants of the newly formed universe develop a history which includes a rationale for their existence, and a mini-God to oversee this mini-Infinity.

- There can be local earth Gods and local earth Infinities for Catholics, Muslims, Jews, etc., but there must also be a major, more universal God for our universe.

PART III

Lifespan and Factors Affecting It: Humans

CHAPTER 5

Longevity in Living Systems, Theories and Speculation

How old are you?
How do you know?
Who is older, you or I?
How old are your lungs? Your heart? Eyes? Skin? Hair?
There is a difference, you know.

And a variability of lifespan. Man versus woman, between animals, trees, corporations and civilizations.

Can you hear time flying? Listen to the silence. Do you hear the clock ticking away, tolling the beginning, the middle and the end? Hey, rabbit in my garden, do you know that you will die in a year or two? That your lifespan is clearly circumscribed? Do you ever talk of these things with your brothers and sisters as you munch on the sweet green leaves which taste so good?

Everyone gets nervous when we talk of aging. Don't tell me I'm going to die. I won't hear of it. But something is going on. Changes are occurring so at eighty your heart can pump blood only at about half of what it could do at age sixty. Body temperature is down. But the Bristlecone Pine tree can live to 4,000 years; some even to 4,600 years. A fly may live a day; a flea, 30 days; a white rat, 4 years; dogs and cats, 20 years; chimpanzees and horses, 40 years; hippopotamuses, 50 years; Indian elephants, 80 years; fresh water mussels and some fish, 100 years; large tortoises, 150 years.

Cro-Magnum man believed a cave painting held the soul of the subject. Aztec souls went to the sun. Borneo natives inferred seven souls in every body; that these fly from us at death through the big toe in the shape of a butterfly. Buddhists had an Eightfold Path of Righteous Living: death and life and death, again and again, until we wend our way through a long chain of rebirths to Nirvana. The Muslims prescribe a Heaven of Sensual

Delight. The Greeks poured wine on the fresh graves during burial and sacrificed animals to feed the resident souls — and envisioned the Elysian Fields, located by Homer on the western border of earth. The alchemists considered that the sun ruled the heart, the moon the brain, Jupiter the liver, Saturn the spleen, Mercury the lungs, Mars the bile, Venus the kidneys.

We yearn for immortality. Don't you? At least cannot we have a long life in good health, with our senses and strengths unimpaired, with sufficient money to enjoy the good life? And if we suffer some losses, why cannot we rejuvenate our being, especially our sexual powers and appearance? Throughout history we have sought the magical cure, the extract from vital organs which when ingested or injected, provides the restorative healing capability. Our modern research, though disguised and bearing the cachet of scientific respectability nevertheless continues in the vein of our ancient predecessors, exploring the effects of nutrients or hormones on lifespan. Injecting spleen cells from young mice into old mice can extend longevity by one-third in some cases. Something called antabotone extracted from the squirrel's brain lowers the body temperature of rats about five degrees, apparently slowing the metabolic processes and extending their tenure. Diminishing the fruit fly temperature from 25 °C to 19 °C seems to double its lifespan. Rotifers react similarly. For people, lowering body temperature by two degrees decreases our basal metabolic rate by 8 percent. Conversely, wound healing is improved by raising the temperature; higher temperatures also accelerate the destruction of invading bacteria and viruses by cells called phagocytes. A chemical, monoamine oxidase, in the brain, seems to be related to depression and schizophrenia. Treat these disorders by suppressing the levels of monoamine oxidase? Or is it the immune system which is the cause of our demise: old people die of infection and the complications derived therefrom, say the researchers who seek ways of bolstering the immune surveillance system. Remove the thymus and you decrease the concentration of the hormone, thymosin, and retard the immune system. A lack of suppressor cells is thought to allow an uncontrolled antibody production. This may or may not be advantageous for us; on the one hand we need the antibodies to fight disease, yet on the other hand these excesses may lead to deranged antibodies which attack our own cells, a masochist's dream, resulting in the autoimmune diseases of old age such as lupus erythematosis.

None of the foregoing ideas and observations provide the answer to the recondite question as to the cause of aging, but considering them in toto, perhaps we can surround the unknown, this black hole of knowledge — where everything we do, all questions asked, get devoured; where no solutions emerge. One favorite theory of aging, the free radical theory, says that there are certain parts of molecules — free radicals — which are ephemeral, transitory, very unstable, highly reactive and therefore eminently disruptive to the chemistry of life in our bodies. They enter into our vital oxygen reactions, form side products which are unneeded or noxious, weaken cell membranes, disturb the delicate DNA machinery and in general produce weakened cells and organs. Radiation exposure as well as oxygen and ozone are thought to be free radical generators. The animal or person exposed to radiation suffers the apparent effects of an accelerated aging process: thickened artery linings; wrinkled skin; and accumulation of so-called old age pigments such as lipofuscin and amyloid. A small animal can be killed by high concentrations of ozone; the astronauts in the early days of the U.S space program suffered eye problems when exposed to a pure oxygen environment. We are asked to eat more unsaturated (soft) fats in order to prevent heart disease yet we now know that free radicals are more easily generated by the unsaturated fats, as opposed to the hard, saturated variety. Hence the dilemma: if all this is true, if by the very nature of living, in the course of normal existence, we generate these harmful free radicals, then first of all why do we humans tend to live so long and secondly cannot we do something to neutralize the free radicals? The explanations offered are glib, unproved and provocative. We live our lives, we die our deaths in a controlled manner; the rate of living is moderated by naturally occurring free radical scavengers residing within the cells of our bodies and concomitantly we ingest foods containing these same free radical neutralizers (also called antioxidants). Vitamin E is one of them, so are the vitamins A and C as well as a class of chemicals called mercaptans and others with abstruse nomenclature — BHT, 2-MEA, Santoquin. In the diets of rats, these free radical scavengers (antioxidants) enabled an experimental group to outlive the control. If true, if free radicals are the bane of our existence, if we can slow their formation, if we can demonstrate that by diminishing the presence of copper (the catalytic agent of the free radical reaction) we live longer, then will you become a believer?

Collagen is a protein material which forms the structural component of our bodies and also the surfaces through which diffuse the enzymes, vital oxygen and other sources of life. In its youthful condition, collagen is soluble in some fluids and appears to be fibrous or stringy under the microscope. With time, young collagen becomes old collagen and insoluble, cross-linked and stiffer than before. And so, you and I as we age, get stiffer, more prone to pull an Achilles tendon if not "warmed up" or stretched sufficiently before exercising. We get less oxygen through our lung surfaces when old for these membranes are mostly collagen; it also becomes more difficult to take a deep breath. Literally painful to do. And so we slow down, stiffen, confined as we are in an ever tightening straightjacket of old collagen, unable to inhale our optimum amounts of oxygen, our vitality diminished, ripe for the other debilitating scourges of old age. A professional football player a few years ago, a quarterback, who at age 39 (advanced for this sport), with a history of lackadaisical attention to physical conditioning, walked onto the practice field for the first time in a new season, threw one football and tore his Achilles tendon. Out for the season and at the end of his career, our star was a victim of old, cross-linked collagen. If you believe this collagen or cross-linking theory of aging, then you may ask, "Where do we find the cure?" What we seek is the magic bullet, a bacterium or virus inclined to devour old collagen. We know they exist for they work quite effectively in soils when our bodies are finally laid to rest there. But there must be certain specificity for it is only the cross-linked, insoluble, inflexible collagen we wish to remove, thereby providing the space for our bodies to lay down a nascent mesh of young, soluble, flexible fibrils. In the meantime, until the panacea is found, we must be content to contemplate the tantalizing, incomplete research results which seem to indicate that a free radical neutralizer such as vitamin C is effective in ameliorating the effects of cross-linking of collagen. Indeed there is some evidence suggesting that old people lose their teeth not so much through tooth decay, but because of shrinking gums (composed of collagen) and the subsequent exposure of tooth roots which yield ultimately a loosening and detachment of the teeth from their bases. Vitamin C is a factor in preventing the cross-linking and shrinking of the gums and associated tooth loss.

The wear and tear (rate of living) theory of aging expresses the conviction that just as the machine wears out, the living system does also. (Survival curves for automobiles, cockroaches and people are similarly shaped.) Gasoline fuels the engine, food energizes the body. There is a certain prescribed duration of light which can emanate from a light bulb in its lifetime (the rating is prominently displayed on the container); shouldn't we expect the same limitation for an animate system in terms of its complement of energy, enzymes or other vital matters and cannot we project a longevity in proportion to the levels of these resources at birth and the rate at which they are consumed? Finally, ultimately, couldn't we expect senile death to approach when the energy or enzymes are depleted or reach a critically low level of concentration? This then is the essence of the wear and tear or rate of living theory of aging. Aided and abetted by fascinating empirical observations, the apostles of this cause are firm in their convictions. For example, in experiments done years ago and verified readily since then, rats were underfed — healthy diets but lacking in the caloric content of the control group — and when the life spans were totaled, these underfed rats lived longer. The feeding regimen could be altered; one day of fasting in a three-day cycle didn't change the results. In different experiments, the heartbeats were summed for rats and elephants. The rats, living two or three years, with heart rates of about 500 beats per minute generated approximately one billion beats in an average lifetime; the elephants, 25 beats per minute, living to about 80 years of age also averaged approximately one billion heartbeats. And then there are cell doubling experiments: cells taken from the youngest human, living fetal tissue, could double around fifty times before the cell population senesced and died. Obtaining cells from a mature person and performing the same doubling experiments yielded only twenty doublings before the population dissipated itself. Indeed the experiment could be refined by removing those youngest cells from their milieu after ten doublings, freezing them in a liquid nitrogen bath, distributing them around the world to other laboratories, waiting six years, thawing the cells and resuming the protocol. Those youngest cells, with an initial potential of fifty doublings, having already doubled ten times, after six years resumed their splitting for about forty more times. Curiously, cells from subjects with cystic fibrosis exhibited normal population doublings in early passages. However, after about thirteen doublings these cells began to double more slowly than the control group and ceased doubling after nineteen compared to twenty-seven for the control. It may be that these cells of cystic fibrosis patients were demonstrating some sort of premature aging syndrome.

These then are the essential theories of aging. Aided and abetted by tangential but no less important new developments related to the control processes within the brain, we are beginning to understand the grand ensemble. The evolving immune system function with age is an example of the control thought to be exercised by the brain. The brain is the origin, from which is derived an understanding of the hypothalamus (governs hunger, satiety, body temperature, water balance, blood pressure, heart rate), the pituitary (influences metabolism, growth, reproduction), the thyroid (affects metabolism, oxygen consumption), and the thymus (generates the essence of the immune system). All of these functions are dependent upon neurotransmitters such as dopamine, serotonin and acetylcholine, affecting the transmission of nerve signals. The level of dopamine in the brain is known to diminish with age, a fact uncovered when attention was riveted on it as a treatment for Parkinson's disease. Found in such foods as wheat germ and velvet beans, L-dopa becomes dopamine in our bodies. It affects the aging process or the symptoms of aging in ways not yet clearly specified except that we know its absence affects hypothalamus function. The thyroid in the neck area has been the focus of attention for half a century or more, in the study of conditions associated with the underproduction or overproduction of the hormone thyroxin which is secreted by the thyroid. Too little thyroxin diffusing into the cells causes the basal metabolic rate to be insufficient for proper energy metabolism. The patient seems listless and de-enervated. Too high a level of thyroxin revs up the cells, developing a hyperactive and hence inefficient operation. Hypothyroidic persons tend to have too many wrinkles, grey hair, low resistance to disease, retarded growth and weakened muscle and cardiovascular function. Hyperthyroid cases exhibit nervousness, decreased growth, and shortened lifespan. Old persons in good health generally show no abnormally high or low levels of thyroxin in their blood. This was a seeming paradox until it was established that the problem was not concentrations in the blood but the inability of the thyroxin to penetrate the enveloping cell membrane. Even with normal thyroid function and acceptable levels of thyroxin in the blood, this energizing hormone thyroxin is apparently unable to diffuse into the cells in proper amounts. The result is a rundown system, not operating at its proper level, contributing to the common observation that old folks' basal metabolic rates drop steadily with age, their vitality is diminished monotonically and they become more prone to be victims of killing stresses such as disease, physical injury, psychological traumas and other vectors younger persons are able to shrug off. There is some evidence that this increasing inability to get thyroxin into the cells is a programmed event, orchestrated by the pituitary. Something is excreted from the pituitary in our mature and later phases of life which seems to block the movement of thyroxin. Ominously called the "death hormone", its presence (though not its identification) has been indicated indirectly. By excising the pituitary of some old rats and then adding thyroxin to the bloodstream, the debilitating effects of a thyroxin lack are reversed; the rats become seemingly rejuvenated. Find a way to neutralize this so-called death hormone, find a way to facilely get thyroxin into the cells of old folks and will you create a revolution? Remember the discussion of underfeeding and its effect on lifespan? Suppose underfeeding maintains the pituitary in a youthful condition, delaying the time when, if you believe those who say these events are programmed, the death hormone is released.

The acolytes of the immune theory of aging focus on the thymus. Here is located the essence of the immune system, they believe. The thymus receives cells from the bone marrow (B or bone marrow derived cells) and through action with the hormone thymosin secreted by the thymus, causes the development of T or thymus derived cells. The T cells enter the blood and lymphoid tissue to become killer cells (lymphocytes) capable of attacking cancer cells, viruses and bacteria. B cells also enter lymphoid tissue. They secrete antibody molecules which provide immunity against disease causing germs. The thymus atrophies at an early age, shortly after puberty in humans. B cell function declines shortly thereafter. T cells seem to change most with age, exhibiting early difficulty in dividing. Their loss of function is believed to affect B cell antibody secretion, leading to an increased presence of deranged antibodies called autoantibodies which attack our own body's tissue. Old people have elevated levels of autoantibodies, and suffer from what are believed to be autoimmune diseases: rheumatoid arthritis and maturity onset diabetes.

Harrison examined parabiosis (the joining of the blood circulatory systems of two animals). Parabiosis did not rejuvenate old mice, and suggested something within the old organism had a negative effect upon the younger mice. Tolmosoff addressed the free radical theory of aging and suggested the superoxide free radical as the debilitating influence. Attention has focused recently on superoxide dismutase (SOD), the neutralizer of the superoxide free radical. SOD may remove the superoxide free radical. SOD's presence has been established in the liver, brain, the heart of rodents and primates including man. We humans seem to have the highest levels of SOD. Longer-lived species may have a higher degree of SOD protection against the byproducts of the oxygen metabolism which yield superoxide free radicals. Tolmosoff also suggested a maximum lifespan potential for us of about 110 years. He reported maximum lifespan energy expenditures of 200 to 300 kilocalories per gram of body weight for many mammals; primates in general generate two times this amount and man, about three times as much.

The original questions still apply. How old are you? How do you know who is older, you or I? How old are your lungs, heart, etc.? We know our maximum work output declines at a greater rate compared to the individual organs of our body. So the question becomes important? How old are you? What do you mean by the question? The whole may be older than the individual parts. In crank turning experiments, we suffer a greater loss as a whole than do the involved muscles. Similar conclusions for maximum breathing capacity vis-à-vis isolated lung function. Apparently the insults delivered to our organs have a cumulative affect, so that finally we see a reduction in our ability to tolerate sugars, a condition derived from a slightly lowered ability of the pancreas to respond to blood glucose concentrations. The unused portion of the lung volume increases with age in proportions beyond the range of change in collagen.

Old people in the Caucasus Mountains between the Black Sea and the Caspian Sea have been the object of much scientific and commercial interest (as yogurt salesman). Another pocket of long-livers resides in the village of Vilcambia in the Ecuadorian Andes, a third settlement is found in the high valley of the Karakorum Mountains of northernmost Kashmir, at the extreme western end of the Himalayas. We on the outside are naturally curious about the common denominators distinctive to these geriatric ensembles, people who live long lives in good health. We notice that the populations of these lands do hard physical labor. Their diets are low in calories. They continue to be active and respected, though aged — which seems so different from our modern, sophisticated societies. On closer scrutiny, however, some are not as old as they claim. It appears there was considerable falsification of birth dates and ages under the Czars and afterward, done for a number of reasons, including an attempt to avoid military service. During Stalin's time the Russians were quite anxious to please Stalin, a former resident of one of the pockets of long-livers. By association he could claim to have an extended life potential. Brown-Sequard in 1889 believed our aging gonads were the key: testicular extracts should have a rejuvenation effect. By the 1920's Serge Vornoff grafted sex glands from young non-human primates into aging humans. At the turn of the twentieth century Elie Metchnikoff, an associate of Louis Pasteur in Paris, theorized that aging was due to the effects of intestinal toxins. He advocated a diet rich in lactobacillus bulgaris (a bacterium in yogurt). A. A. Bogomolets invoked the collagen theory and found some sort of serum for revitalization of our connective tissue. Anna Aslan, in Romania, used Gerovital H3, a procaine material like the dentist's Novocain, for treatment whose aim is rejuvenation. Hans Selye claimed stress as the aging vector. Background radiation was the culprit according to physicist Leo Szilard. Szilard suggested that radiation randomly impinged on our chromosomes, inducing the changes which produce a steadily weakened organism. Leslie Orgel authored an error or catastrophe theory, whereby the DNA protein synthesizing apparatus is damaged, yielding faulty templates for protein production. John Bjorksten in Wisconsin was the major proponent of the collagen or cross-linking theory of aging. C. M. McCay at Cornell, prolonged the life of rats up to 25 percent by diet restriction in early life.

Max Rubner in 1908 totaled the number of calories consumed per gram of body weight during the lifespan of mammals and found the amount was approximately the same for the mouse (3.5 year lifespan) and the elephant (70 year lifespan). Cumulative heartbeats over these lifespan were amazingly similar — 1.1 billion and 1.0 billion respectively. The mouse, living so rapidly for so short a duration, and the dissimilar elephant — both have a rhythm of life in phase. Live faster, live less. Raymond Pearl at Johns Hopkins working with fruit flies during the early twentieth century found the environmental temperature to be a factor affecting lifespan. Raise the temperature, shorten the lifespan and vice versa. Women live about eight years longer than men; perhaps because they have basal metabolic rates about six percent below men. Bernard Strehler at the University of Southern California discovered that mice with depressed body temperatures live up to twice as long as others. A. V. Everett believed in a biological clock or pacemaker, as did Caleb Finch at the University of Southern California. W. D. Denkla at the Harvard School of Medicine proposed the "death hormone", secreted from the pituitary, which controls the thyroid and the thymus and hence the aging process. Leonard Hayflick's *in vitro* experiments demonstrated a limited doubling potential lifespan for human diploid cells.

The lexicon of researchers in gerontology includes Denham Harmon who demonstrated a life extension possibility by feeding mice 2-MEA (2-mercaptoethylamine), BHT (butylated hydroxytoluene), vitamin E and Santoquin (a quinoline derivative) — all of them free radical neutralizers or scavengers (antioxidants). Mothers fed antioxidants had long-lived offspring. Charles Barrows increased the longevity of rats by about 30 percent by halving the protein content in the feed of mature rats. Allan Goldstein of the University of Texas in Galveston injected thymosin into rats and boosted their immune function. Paul Chretien successfully used thymosin in lung cancer therapy in conjunction with chemotherapy. Others applied anti-aging concepts such as temperature-lowering drugs, or experimented with L-dopa or collagen cross-linking inhibitors (vitamin C) or cell membrane stabilizers (selenium) or lipofuscin scavengers (antioxidants).

The octopus dies after spawning, as does the salmon. For the octopus, corticosteroid hormones are released by the optic nerve. Remove the optic nerve and you can double the lifespan of the female octopus. Harman claims a diet containing free radical scavengers (antioxidants) has increased the lifespan of mice, rats and fruit flies, inhibited some forms of cancer, allowed less deposition of the yellow old age pigment, amyloid, and increased the immune response. The proper diet can add 5 to 10 years of life, he claimed. The damage inflicted by free radical reactions are: (1) cross-linking of collagen and elastin, the structural protein of the body; (2) breakdown of the membrane and cell material through which oxygen, enzymes and other vital substances diffuse; (3) accumulation of aging pigments by polymerization of lipids and proteins; (4) changes in mitochondria membranes by excess oxidation (peroxidation) and (5) thickening of the small blood vessel walls (fibroisis) by oxidation of the blood serum and the wall linings. Thinner people should live longer with the onset of old age disease delayed. Easily oxidized amino acids such as histidine and lysine increase free radical reactions and shorten lifespan. Radiation is a free radical generator, causing tumors in mice. The radiation effect is enhanced when unsaturated fats are added to the diet. The incidence of breast cancer in women seems to increase in proportion to the unsaturated fats in the diet. However, tumor incidence of rats fed high levels of safflower oil (an unsaturated fat) can be diminished when vitamin E (an antioxidant, free radical neutralizer) is added to their diets.

The ancient Egyptians looked for a longevity elixir. In pre-biblical time the Chinese saw aging intermeshed with lifestyle (Yin and Yang). The Greek, Hippocrates, said aging was caused by a decrease in body heat. Galen proposed changes in body humors — increased dryness and coldness — as the driving force. The philosopher Maimonides considered life predetermined. Roger Bacon proposed to stop the aging process by good hygiene. Da Vinci performed autopsies; Francis Bacon looked for our vital spirit. Santorio promulgated a very modern-sounding theory of hardening of the fibers and the progressive consolidation of earthy materials within the body. Darwin linked aging to a failure to respond to the sense of irritability of the tissue (perhaps another way of referring to the immune system). Warthin felt we lost energy. Hufeland told again of a vital force. Metchnikoff suggested autointoxication; we must at all costs prevent intestinal purification. Durang and Fardel encyclopedically described changes in the body with age. Charcot applied physiological principles and tests to study the changes. Canstatt showed cells die and are not replaced. Minot at the turn of the century studied mortality rates, following in the tradition of Gompertz. Kohn in 1963 said if cancer could be cured, lifespan could be extended by 3 years; stop cancer and cardiovascular diseases and we can add 10 to 15 years. Comfort reported that our ultimate lifespan, the maximum potential, is a human characteristic independent of culture.

The maximum survival of humans is estimated to be around 115 years; the maximum doubling potential of our cells seems to be about 50 times. Consider the relationship between the two: a mouse, 3 years lifespan, 12 doublings; chickens, 30 years, 25 doublings; Galapogos tortoise, 200 years, 140 doublings. Is your ultimate lifespan reflected in the doubling capacity of our cells? Is there another circumscribing factor, such as the accumulation of metabolic detritus, lipofuscin, which can be used as a benchmark? We know lipofuscin is found in the cytoplasm of the cells in the brain, muscle, myocardium, adrenal cortex, testis, ovaries, liver and kidney. Centrophenoxine is capable of stimulating the metabolism of nerve cells and is able to cause lipofuscin to disappear from nerve cells and myocardium. Animals treated with centrophenoxine lived longer than a control group and had a better learning capacity. One part of centrophenoxine is a precursor of acetylcholine which comes from choline, a neutrotransmitter. Should we combine centrophenoxine and SOD therapies? Would we get a synergistic effect? We know that 80 percent of the superoxide free radical in the heart is trapped by SOD. The remaining 20 percent reacts with membranes, inflicting its biochemical hurts.

Hayflick removed the nucleus from the cell cytoplasm of both old and young cells. He placed the old nucleus in the cytoplasm of young cells and found the young cells then doubled according to the older cell's lifespan potential. Conversely, a young nucleus in old cytoplasm yields a doubling and hence living capacity controlled by the young nucleus. Maites focused on the loss of reproductive capacity and imputed this loss to a deficiency of amines such as dopamine in the hypothalamus. Ovulation in old female rats can be induced by epinephrine, from a class of chemicals called catecholamines produced in the hypothalamus. Put ovaries from old rats in the young and the ovaries being cycling. Conversely, young ovaries in old rats: everything ceases. Seventy percent of an old rat's liver can be removed and it will regenerate; the new tissue is old and responds weakly to glucose. Even though we've achieved regeneration of the liver, the DNA coding is sensitive to the fact that it is an old system giving the orders. In the 1890's thyroxin was prescribed for humans with indolent thyroids. The reversals were dramatic: wrinkles disappeared, grey hair turned black, brown or blond again; resistance to disease returned to normal. However for those who were just old, with normal thyroid function the treatment didn't work. Some favor a stochastic philosophy which says life is a series of events governed by the laws of probability. By virtue of cells and molecules bumping into each other, changes occur; changes leading to free radicals, wear and tear, errors in the synthesis of DNA and weakened immune vitality. When cells double, what is produced in not two identical, equally youthful daughter cells, but two different daughter cells: one able to continue to proliferate (uncommitted) and the other is unable to do this and hence committed to death. These theories have some utility in attempting to understand why epithelial cells of mouse duodenum and small intestine have a turnover rate of two days, parenchymal cells of the liver are replaced in 160 to 400 days, nerve and muscle cells live for 1000 days and never divide.

Bee stings or ointment containing bee venom somehow are helpful in stimulating the immune system and are used in the treatment of rheumatism, infectious polyarthritis, spondyl-arthritis deformans, trophic ulcers, asthma, and migraine. The venom can be delivered topically in oily or aqueous solution or by subcutaneous injection. Considerable swelling and inflammation results, without fatalities. And it seems to work, not only for some old patients but also for a 15 year old girl who suffered from precocious or premature aging. She had low levels of active T cells and had a family history of cancer. Can we use bee sting therapy to treat lupus erythematosis, an autoimmune disease?

You can increase lifespan for mice and guinea pigs by adding the amino acid cysteine to the diet, or a combination of thiazolidinercarboxylic acid and folic acid (all containing sulfur and hydrogen atoms chemically bound). Passwater and Welker theorized that antioxidants (free radical scavengers) in the diet could add 5 to 10 years to our lifespan; sufhydryl compounds (containing sulfur) which afford radiation protection and are free radical scavengers could add another 2 to 5 years and protein missynthesis correctors (containing selenium) could contribute 5 to 10 additional years. Taken together in one diet, assuming the three factors act synergistically according to Passer and Welker, an extra 30 to 40 years of longevity can be obtained. The sulfhydryl compounds taken in conjunction with vitamin E are supposed to increase our tolerance to selenium which is supposed to be able to control errors of DNA production of protein. Foods containing selenium include tuna, herring, menhaden, anchovette, brewers yeast, wheat germ, bran, broccoli, onions, cabbage and tomatoes. Foods high in sulfur amino acids are eggs, cabbage, muscle meat, wheat germ oil, leafy vegetables, fish, whole wheat, vegetable oils.

Ancient Egyptians ate garlic to ward off death. The alchemists of China prescribed gold and mercury to do the same. People have eaten mandrake roots and allowed themselves to contact fever blisters, to get monkey gland transplants, to inject procaine all in the search of good health and rejuvenation. Are we programmed to self-destruct with time, a basically pessimistic theory of aging? Are there errors constantly cropping up, or as in the masochist's dream do we misidentify ourselves, see our reflection as a foreign invader and attack until there is nothing left to recognize? Are the chemical bridges of collagen and other structural materials altered with age, causing us to function dolorously — increasingly so until we are sufficiently weakened and die? Or do we in living, desecrate our living edifice with biological detritus? We search and research the means for survival, from vitamins E and C to bacteria which can eat old collage, to thymosin and its effect on the immune system, to centrophenoxine to stimulate the hypothalamus, to dopamine for proper brain function and even to starvation for a slower growth rate.

If it were possible to protect living things from all possibilities of unnatural death and provide for complete nutrition and a controlled environment, then, ideally, the survival curves would be similar to that obtained for rotifers. These rotifers live to the fullest extent of their biological makeup and reach senescense at around 30 days. At this age they lose their vitality or ability to survive and die off quickly as a group. These rotifers are bred to be genetically similar, with uniform characteristics and potential. Thus their uniform demise is not what might be expected in humans.

We know a dog does not live as long as a human; we know an elephant can live longer than a rabbit. So as we ask questions about how long we can expect to live, we need to ask additional questions pertaining to how long one species will live when compared to another species. And we must discuss the expected lifespan for those who lead "normal" lives as compared to those who suffer premature deaths due to accidents. Obviously the animals in wild are more vulnerable to early death. So there are many survival curves, for each species, all with more or less the same shape, but terminating at different age levels. Few deaths occur early in life for the healthy mice reared very carefully in captivity, and so the death rate is very low. Then as the mice reach maturity and old age (about 500 days) they begin to die off more quickly. At 900 days almost no survivors remain.

Apparently trees also age, though very slowly — much more slowly than the "living" animals. So even trees are not immortal. (Everything seems to die eventually, though the lifespan vary.) Mortal man at 85 years of age has only about 10 percent of his cohorts with him; only 10 percent of the original sample is still alive. For the thoroughbred horse, the 10 percent level is reached at 27 years of age; for the laboratory mouse it is one-and-one half to three years, depending on the strain of mouse.

For dogs, the survival curves show that the lifespan of large breeds is considerably shorter than that for smaller breeds. Some researchers have analyzed survival curves for mammals and dogs, coming up with empirical equations which relate lifespan to body weight.

For those of us subjected to a wide range of climatic conditions, individuals from colder locations seem to be longer-lived and slower-growing than those from warm places. For example, some mollusks which were placed in jars at 100 days of age lived for 125 more days at 23 °C, while at 7 °C to 20 °C; the additional lifespan was 280 days. In experiments with rotifers, lowering the temperature 10 °C could prolong their lifespan about four times its normal expectancy. Cutting in half their food intake could extend life threefold.

This type of information suggests that longevity correlations need to be qualified, bringing other parameters into the description. There are relationships between lifespan and diet, body surface area, weight, metabolic rate, and dependence has even been found between brain size and lifespan. Birds live longer than mammals of comparable size. Cold-blooded animals live longer than birds or mammals, and it is suspected that these cold-blooded creatures have a rate of aging that is sharply temperature dependent.

It is interesting to observe that there seems to be an optimum temperature for maximum lifespan of the water flea. Depart in either direction from this optimum and the lifespan is diminished significantly. For trout and the drosophila fly, raising the temperature of their milieu increases their metabolic rate and shortens their lives. If I overeat and am constantly well-fed, will my rate of living be higher than usual; and will I therefore live a shorter life than others? (Overfed house flies are shorter lived than a control group which was fed normally.) Does it matter how long it takes me to reach full size; and if it does, can I alter my growth rate by control of my diet? (Small apes take three years to reach full size and live to 10 years of age, while the chimpanzee attains his maximum size at 11 years of age and lives to 40.) The ratio of lifespan to age at full size is about three or four to one. Slow the growth, lengthen the lifespan? If a woman is fertile, will she outlive others? (For roaches, higher fertility females have shorter lives than virgin females who lived up to 50 percent longer.) If we are crowded together, as we are in many cities, will we have a shorter lifespan? The answer is yes for the drosophila fly. Is it true for us? If I live where the oxygen content in the air is low, can that affect my lifespan? Apparently yes for some flies where a pure oxygen environment apparently speeds up the rate of living and shortens life. Does this information on flies apply to us in our polluted air? The cynical answer is that the oxygen concentration in the polluted air probably will not be a major factor affecting our duration on this earth — other things in the air will probably do a better job of killing us.

The metabolic rate drops with age, as does the conduction velocity of some nerves. After age 30, some researchers estimate that we suffer a loss of function of about 1 percent each year. After age 40, the human death rate doubles each successive eight-year period.

There may be something more fundamental operating on living things, something which may govern the aging process and ultimately be able to define life. Organisms can be considered islands of orderly structure; if this is true, then aging may be an increase in randomness or disorderliness. A failure of the buffering power of the living system to control the upsets encountered may scramble the information stored in the chromosomes. Some say that this disorganized information (noise) can accumulate in the cellular information system. Stresses that scramble the chromosome information on this cellular level, such as radiation, seem to accelerate the aging process. Background radiation we all encounter on earth may be a factor in aging. This hypothesis has fascinating possibilities, suggesting experiments where whole animals are raised in radiation-free environments for a number of generations.

Talk of randomness and information content leads to the concept of entropy, which can be defined as a measure of the disorder of the system. This line of reasoning says that the force of mortality is dependent not only on time, but also on some physiological equivalent to the wearing out of an automobile (even though we continue to replace its parts as it wears out). In other words, there may be some innate property of the whole organism related to the disorder (entropy) generated by the organism which is responsible ultimately for the death of the living system. Death may come when the level of this generated disorder reaches some critical mark; or it may be a cumulative thing, with death occurring when the summed total of this disorder or entropy achieves another critical level.

CHAPTER 6

Energy, Entropy, Basal Metabolism and Lifespan

Life is a series of chemical reactions accelerated or moderated and modulated by enzymes. We store or expend energy by this route and if we are constant temperature beings — as we are — then we transfer to the surroundings most of the resulting energy as heat, generated by the metabolism of fats, carbohydrates and protein. Fats are insoluble, transported in the blood, bound to protein as lipoprotein. They may be triglycerides, derived from fatty acids and glycerol. Fats in foodstuffs and fatty deposits in animals consist of mixtures of triglycerides. Some typical fatty acids are palmitic and stearic acids (saturated or solid fatty acids) and the unsaturated or liquid fatty acids such as linoleic acid (found in vegetable oils) and arachidonic acid (in fish and animals). Fats, when absorbed, circulate to the liver where as triglycerides they are hectored into becoming fatty acids. From the liver fats pass into the blood and then to adipose tissue for storage. Blood triglyceride levels are increased by fats high in saturated fatty acids such as beef fat and butter. Thyroid hormones and physical exercise diminish their presence. From adipose tissue fats are carried to muscle where they are oxidized to carbon dioxide and water to yield energy. Fat utilization is enhanced by adrenalin and noradrenalin and growth hormone from the pituitary. Emotional stimuli increase fat mobilization.

Sugars and starches (carbohydrates) are absorbed as glucose. Insulin secretion is stimulated by glucose; glucose moves via the bloodstream to nerve tissue where it is oxidized to release the energy required for neuron function. And to muscle, to be stored as glycogen or oxidized to produce energy or converted into fat when in excess. Glycogen can be reactivated by conversion to glucose by adrenalin. If done rapidly without sufficient oxygen (anaerobically), lactic acid is produced. In starvation, when no carbohydrate is available, the energy of life comes from fat, changed in the liver to ketone bodies, acetone and other substances, most of which are eliminated in the urine. Insulin facilitates the passage of glucose through the cell wall. Adrenalin produces a rise in blood glucose by releasing it from the liver. Caffeine raises the blood glucose level. Protein can be oxidized in the liver for energy (producing urea).

In our cells immediate energy flows from the breakdown of adenosine triphosphate (ATP) to adenosine diphosphate (ADP) by the reaction of glucose and ATP to yield glucose-6-phosphate plus ADP. Heat is liberated, serving to maintain normal body temperature. Exercising after a meal will approximately double our body's heat generation (measured as the basal metabolic rate, the heat transferred from the body to the ambient air under carefully controlled resting conditions). The metabolic rate may increase in response to a cold environment, on the order of 10 to 30 percent.

Metabolizing vegetable and fish oils (containing polyunsaturated fatty acids) effectively lowers cholesterol levels; the opposite is true for animal fats (saturated fatty acids). Large meals lead to higher levels of fatty materials in the blood and have the effect of boosting the net caloric content in the food. One large meal in the evening can be devastating in its effect. Basal metabolic rate (BMR) measurements seem to vary by 10 percent during the day. Overfeeding can raise the BMR by 29 percent; underfeeding can drop it by 17 percent. The BMR is defined as the energy output of an individual under standardized resting conditions: bodily and mentally at rest; 12 to 18 hours after a meal; in a thermally neutral environment. Circadian rhythms affect the basal metabolic rate and should also be factored into the experimental conditions arranged when doing BMR measurements. Minor disturbances in the testing room can cause alarums and an 11 percent change in the BMR. The effect of changes in deep body temperature on metabolic rate is quite well established: a 1°C rise caused by fever or other means translates into a 12 percent increase in the BMR. Similarly, the effect of hypothermia (lowering the body temperature) in reducing the metabolic rate is well known; use is made of this in some surgical procedures. The basal metabolic rate decreases during undernutrition, to a greater extent than mere loss of body weight can explain; resumption of feeding restores the metabolic rate to normal long before body weight achieves its initial value. Should the subject again be underfed, the metabolic rate now falls more rapidly than before. High protein diets have a thermal effect, that is, they raise the BMR; fats and carbohydrates do also, but to a lesser degree. In obese persons on diets, there is a more rapid loss of weight if they are fed their equivalent calories in smaller, more frequent doses. As obese people lose weight their BMR drops; they may have BMR values lower than the rest of the population and therefore, even while eating normally may continue to maintain an overweight condition.

BMR values tend to reach a minimum during the summer months and a maximum in the winter for some Japanese and Korean men. Energy expenditures at rest and during physical exercise in a thermally neutral environment are greater in winter than in summer. Birds and mammals maintain constant body temperature by metabolic means. Usually our body temperatures are set above ambient conditions. On the other hand, reptiles and fish assume the temperature of the environment, their metabolic processes just sufficient to maintain life but not able to fix an isothermal body temperature. We (mammals and birds) require 5 to 10 times more energy to live than reptiles of similar size and body temperature.

In examining the question of how long we live, we need to face one misconception. Figures are bandied about, as to the length of life of the human species. We speak in terms of life expectancy and report that for males this has climbed to the seventies; for females life expectancy is in the high seventies. These statements are misleading for the numbers represent the average life expectancy, taking into account the early deaths in infancy, by accident and the myriad of other causes. Naturally life expectancy increases as medicine progresses and we find more ways to maintain the life of infants and cure diseases of the young. There is another angle to this, in that we can calculate the additional years of expected life, as a function of the age already achieved. For example, at age 55, how many years of life should you expect to achieve, based on the data already available on those 55 and over? The answer may surprise you: we can expect to live way beyond the 70 years that's been quoted so liberally. What we seek is the ultimate lifespan, that age we would achieve if we proceeded to a senile death, uncomplicated by sudden death or catastrophic illness.

As we get older, 60, 65, 70, 75, 80, 85 . . . , the additional years of life get smaller and smaller, but at the same time life expectancy reaches higher and higher values, aiming for the ultimate life expectancy, that age where additional years shrinks to nothing. In other words, extrapolating from 65 to beyond age 85, we seek that age for males where additional years go from 13.7 to 10.9, 8.6, 6.8, 5.3 . . . , to 0.0. We discover by this procedure that the ultimate lifespan for males is 103 years and 110 years for females. Faced with these numbers, we need to develop a renascent view of age, especially old age. It would seem that middle age should now include the 60's, and then we evolve as young-old, middle-old and finally old-old (about age 90). We seek the knowledge to enable our population to remain in essentially good health into young-old age. Thereafter the deracinations of middle-old age become pronounced and it becomes simply a matter of keeping the old-old comfortable and in a decent frame of mind.

There are claims that the lifespan of mammals is related to brain size and the lifetime accumulation of metabolic activity (the total heat evolved or the energy expended in the course of living). Calloway hypothesized that there is a fixed amount of bodily heat generation potential available and this may be 20 kilocalories per kilogram weight per day. With time our vitality diminishes until a critical plateau is reached when there is not sufficient energy available to maintain life. And so we die.

Living systems are highly improbable and complex, composed of subsystems, organs, tissue, cells and processes. The human system involves a consanguity of 60 trillion cells. Constant work and energy expenditures are required to keep them in their proper places. Some imperfections always remain. At first these imperfections are subtle and unnoticeable on a microscopic level; with time, however more and more cells are not restored to their original conditions or configurations after some dislocation or stress. These imperfections gradually accumulate until a weak spot occurs in a vital organ leading ultimately to the collapse of the entire system. The errors induced are also attributed to the effects of entropy, a concept borrowed from thermodynamics. Highly ordered systems carry low entropy and much stored information. Entropy can be defined as a measure of the disorder or randomness of a system engaged in spontaneous actions: it predicts that the system when left to itself inevitably deteriorates with time until its increasing state of disorder achieves a maximum (synonymous with death, the ultimate state of disorder). If the death of an organism is viewed as the state characterized by maximum disorder and if the wear and tear theory offers one reasonable explanation of our longevity, then it should be informative to determine the total entropy production during our lifetimes. Comparisons of the lifetime entropy production of rats, dairy cattle, mules, horses, guinea pigs and humans were estimated; the resulting numbers were quite close, of the same order of magnitude except for humans who were 4 to 5 times higher. From this questions can be asked, such as: Is the lifetime entropy accumulation the same for all animals? Can a new age-scale be defined based on these ideas? Why are humans different?

For living systems, our energy content, entropy production, vitality and rate of living are enmeshed with the chemical reactions within our bodies, affected by temperature. The heat evolved, measured under proper conditions is called the basal metabolic rate (BMR). For each of us there is an internal temperature at which we are comfortable; above this, vasodilatation and active sweating occurs, below the set-point vasoconstriction and increased metabolism due to shivering is called for. Under fasting conditions (no food for 12 to 14 hours) and at rest (subject lying quietly under no overt stress), the energy of living is converted almost entirely into heat rejected to our surrounding environment. Astronauts-men and women-living in a space habitat with gravity at 50 to 75 percent of that on earth exhibited lower metabolic rates for reasons still obscure except perhaps for the fact that less energy was required to properly maintain muscle tone in outer space than on earth where gravity is more of a factor pressing in on us. Cancer patients record higher metabolic rates than normal. Those suffering cardiovascular disease are likewise hypermetabolic. A group of 50 smokers were found to be experiencing greater sleep difficulties than a group of 50 non smokers matched by age and sex. Sleep patterns significantly improved when the smokers abstained. Smoking raises blood pressure, heart rate and fatty acid concentrations, confirming the suspicion that smokers are on some sort of "high", living at an accelerated pace and are physiologically older than their chronological age would indicate. And probably they will suffer shortened lifespan if you believe the wear and tear (rate of living) theory of aging. This effect on the BMR may explain why smokers who kick the habit gain weight. It is true that in abstinence they seek another form of oral gratification and find it in eating. It is also true that their taste buds recover from the noxious chemical contacts which seemed to shrivel and deactivate them. What may be the controlling factor in the rapid weight gain when smokers cease lighting up their cigarettes probably can be understood in relation to the BMR, which is elevated while the subjects are active smokers. The diet of smokers, adequate to maintain a constant weight (when the BMR is high) becomes too rich in calories when the smoking ceases. BMR values drop and now the diet which was sufficient to maintain a fixed girth yields too many calories and unless the ex-smoker reduces the amount of food consumed, he or she might gain weight.

Atwater and Benedict constructed a whole-body oxygen chamber in which men could live and work for days at a time, where their oxygen consumption and hence metabolic rates could be continuously monitored. In 1894 the impact of food, age, and sex, goiter and myxedema on BMR were studied. Basal metabolic rate measurements were of assistance in the diagnosis and treatment of diseases of the thyroid gland. For normal men of the same size and age we can expect a 10 percent variation in BMR determinations. Within the same individual we also find differences: perhaps

5 percent higher in cold versus hot weather and dependent upon the time of day. Children have high BMR values (twice adult levels) as their bodies develop and change rapidly. There is a peak in early life, one to 10 years of age, after which the BMR gradually declines. Per unit weight, embryos exhibit the greatest BMR, apparently a reflection of the work of differentiation. We regulate heat loss from our bodies by evaporation, radiation, and conduction and convection. Should we find ourselves in a hot, humid ambience, these modes of heat transfer may be insufficient to prevent hyperthermia which triggers an aberrant rise in the metabolic rate. When things get too cold, we shiver and thereby raise the metabolic rate. In the lowered pressures of space capsules, BMR values drop by 5 to 20 percent. Persons in the tropics average a 6 percent lower than standard BMR. Caucasians going to the tropics experience a gradual decrease in body heat production. Vegetarians or those on low protein diets have lower metabolic rates than non-vegetarians with high protein consumption.

It is clear enough that life as we know it cannot be maintained even in cold-blooded animals without the production of heat. The amount of heat evolution is naturally greater in warm-blooded creatures whose body temperatures are almost always above that of the surrounding air. One quarter of the heat generated is lost though the evaporation of water from the skin and lungs and consequently must be accounted for in calculating basal metabolic rates. Heart action generates about 4 percent of our metabolic heat: breathing movements about 10 percent: kidney metabolism may be 5 percent; other organ functions total 25 percent. Three quarters of the heat produced under BMR conditions is derived from chemical reactions in the resting tissue, all controlled by the thyroid. Remove the thyroid and you diminish the BMR by 40 percent. The higher metabolism of children may be caused by a growth factor and the onset of puberty. The mouse is short-lived (high BMR), the horse is long-lived (low BMR). Obese men, heavier than 90 percent of their cohort group or more than 15 percent overweight had basal metabolic rates which were less than the average by 13 to 25 percent, raising a question as to whether their obesity might be caused not so much by overeating but by the inability of their bodies to assimilate and "burn" the food efficiently. Conversely, thinner persons (lighter than 90 percent of the rest or more than 15 percent underweight) showed BMR values greater than average by 5 to 7 percent, indicating perhaps a more rapid rate of living.

With fever or hyperthyroidism the patient may become undernourished though consuming a normal diet simply by virtue of the elevated metabolic processes occurring. In fasting our body is selective: the muscles and glandular organs suffer most of the fat and protein loss, the skeleton and nervous system least. Some students with normal diets of 3200–3600 calories per day were placed on a restricted diet of 1400 calories per day for three weeks without reducing their mental or physical activities. As a consequence of this regimen, BMR values dropped 20 percent, pulse rate was down, systolic and diastolic pressures were distinctly lowered.

One of our chief carbohydrate foodstuffs is starch, a polysaccharide. When ingested, our saliva and pancreatic juices convert the starch to glucose, with the chemical formula $C_6H_{12}O_6$. As such it is absorbed through the intestinal wall into the blood and eventually oxidized (reacts with oxygen) in the cells to form carbon dioxide, water and heat. Oxidation — or combustion — liberates heat as will any other fuel. In burning glucose, the volume of carbon dioxide produced is theoretically, exactly equal to the volume of oxygen consumed and we say that the respiratory quotient, R.Q., the ratio of carbon dioxide to oxygen, is equal to 1.0 for carbohydrate oxidation. Fats furnish about a third of our calories normally but a much larger proportion in starvation and diabetes. The fats are broken down into their constituent fatty acids. The metabolism or oxidation of fats with oxygen yields carbon dioxide and water and heat; the respiratory quotient for this is around 0.7, the amount of carbon dioxide liberated for each unit of oxygen consumed. Fats furnished more heat compared to carbohydrate metabolism (in the ratio of 9.3 to 4.1.) Proteins, constituting about 19 percent of the human body are rapidly altered by hydrochloric acid and pepsin into smaller constituents, amino acids, which are absorbed into the blood stream. This is the material for tissue replacement. Amino acids also are oxidized to furnish energy and heat as do carbohydrates and fats. Protein metabolism releases sulfur and nitrogen compounds which are eliminated through the kidneys. Nitrogen excretion in the urine is used as an index of protein metabolism. Infectious diseases associated with toxemia, cancer and other wasting illnesses are accompanied by the destruction of protein (nitrogen excretion exceeds nitrogen food intake). Metabolism of protein with oxygen yields the usual carbon dioxide, water and heat and a respiration quotient of around 0.8. Diabetics have BMR values higher than average, on the order of 20 percent above average, attributed to extra-ordinarily high protein metabolism. Enlargement of the thyroid and growths on it can raise the BMR by 30 percent. Graves' or Basedow's disease is derived from an abnormal secretion of thyroxin from an enlarged thyroid, resulting in hyperthyroidism: high BMR values, manifestations of nervousness, vomiting and diarrhea. On the other hand, a seriously diminished level of thyroxin secretion yields a disease called myxedema (BMR down 40 percent) which is reversible with doses of thyroxin. Malignancies of the thyroid usually will elevate the BMR. Normal basal metabolic rates generally are subsumed within a 15 percent range (above or below) the standard values. Very severe cases of goiter show an increase in BMR of 75 percent above average. Anemia, leukemia and polycythemia simulate hyperthyroidism: BMR up 11 to 90 percent. Other diseases characterized by BMR values elevated beyond 11 percent of normal are: hypertension, endocarditis; myocarditis; pericarditis, malignancies in general, acromegaly; splenic and pernicious anemia, lymphatic and myelogenous leukemia. Diseases yielding less than 11 percent of normal BMR values are: epilepsy; some malignancies; dysphagia; hypopituitarianism. Considering only the thyroid, diseases producing greater than 11 percent BMR readings were goiter and adenomas while the below 11 percent category included myxedema, hypothyroidism and thyroiditis.

Sleeping results in the lowest heat generation; very light, seated activities such as writing and typing may yield BMR values 50 to 100 percent above the minimum sleep levels. Light efforts, playing musical instruments, ironing, slow walking, show three times the sleep levels; moderate efforts, slow swimming, cycling, baseball pitching and tennis give 5 to 6 times the lowest sleep levels; heavy activities, strenuous swimming, rapid bicycle riding, chopping wood, rapid stairs climbing and soccer generate 8 to 9 times sleep level; very heavy work, very rapid cycling, serious basketball playing, very rapid stairs climbing and wrestling yield heat production rates 13 to 16 times the sleeping levels. Over a sustained period swimming the breast stroke at 3 miles per hour seems to be the most strenuous activity reported: 96 times more heat production than sleeping while the butterfly stroke at 3 miles per hour is next at 75 times the sleep value. Maximum work capacity as measured by oxygen consumption reaches its peak in the late teens and twenties, then declines slowly as we age.

In 1678 it was observed that a shrew would die of asphyxiation in a sealed vessel containing a burning candle. In 1770 measurements of the composition of man's inspired and expired air showed changes in oxygen, carbon dioxide and water vapor. A little later heat evolution from living guinea pigs was quantified. By the early 1800's we knew that some sort of combustion of foodstuffs was occurring inside animals. Around 1840 we understood this was oxidation of carbohydrates, fats and protein. The respiration quotient was measured in 1850 and found to vary from 0.62 to 1.04, depending on the foods consumed and the health of the animal. A whole-body calorimeter was constructed in the late 1870's which could maintain a man and monitor his oxygen consumption for long periods of time. Basic work on heats of combustion of various foodstuffs was accomplished in 1894. Human calorimetry became more sophisticated and accurate by 1897.

Metabolism comes from the Greek word meaning the act or process of change; it is defined in terms of the intrinsic energy utilized by living systems, which comes basically from our food supply. Not all of this potential energy is available; what energy we finally capture is generally stored in the form of adenosine triphosphate (ATP) or creatine phosphate (CP).There are thousands of chemical reactions in the various biochemical pathways we call the living process, but we categorize them generally in terms of the oxidation of carbohydrates, fats (or lipids) and protein:

$$C_6H_{12}O_6 + 6O_2 \rightarrow 6CO_2 + 6H_2O \text{ Carbohydrate (glucose)}$$

Respiration quotient $= \dfrac{6}{6} = 1.00$

$$2C_{51}H_{38}O_6 + 146O_2 \rightarrow 102CO_2 + 98H_2O \text{ Fats (tripalmitin)}$$

Respiration quotient $= \dfrac{102}{146} = 0.703$

$$2C_6H_{13}O_2N + 15O_2 \rightarrow 11CO_2 + 11H_2O + CO(NH_2)$$
$$\text{Protein (amino acid leuceine)}$$

Respiration quotient $= \dfrac{11}{15} = 0.734$

These reactions are exothermic; they proceed with the evolution of heat. We find in general that carbohydrates liberate 4.1 kilocalories for each gram oxidized; fats, 9.3 kilocalories per gram; protein, 5.6 kilocalories per gram. Heat production is an indication of our vitality or rate of living. In humans this metabolic rate is about 60 to 70 kilocalories per hour; during violent activity it may achieve a burst of 1,000 to 2,000 kilocalories in an hour. Men have basal metabolic rates (BMR) 5 to 7 percent above women's values. The male sex hormone, testosterone, can hike the BMR by 10 to 15 percent; curiously, the female sex hormone, progesterone, seems to have a negligible effect on the BMR. The somatotrophic (growth) hormone of the pituitary raises the BMR 15 to 20 percent. Epinephrine and norepinephrine secreted by the sympathetic nervous system and the adrenal glands can kite the BMR 60 to 100 percent for short periods. If your BMR is between −15 percent and +20 percent of average, it is considered normal. Between −30 percent and −60 percent we begin to suspect hypothyroidism and a possible lack of thyroxin in the cells. BMR values, +50 percent to +75 percent, are inferred to be signs of hyperthyroidism, an overactive thyroid, too much thyroxin leaking into the cells. Caffeine increases the BMR, 2, 4-dinithrophenol in small quantities ups the BMR by a stupendous 1,000 percent. Eating tends to raise the metabolic rate; it takes 2 to 5 hours for the carbohydrates effect to subside, 7 to 9 hours for fats, 10 to 12 hours for protein. Undernourishment or fasting drops the BMR by 20 to 30 percent. Metabolic heat generation is increased by a departure in either direction from our thermal comfort zone of 26 to 27°C. Persons in highly emotional states may experience basal metabolic rates 20 to 30 percent above normal; sleep on the other hand is characterized by BMR values 10 to 15 percent below average.

Mice with skin cancers, exposed to hyperbaric hydrogen atmospheres (2.5 percent oxygen, 97.5 percent hydrogen, 8 atmospheres pressure, for two weeks) showed a marked regression of the tumors. Here we seem to have two factors involved: (1) an ambience poor in oxygen, which should be helpful in minimizing the free radical reactions believed to be deleterious to our longevity and immune systems and hence a negative influence on cancer growth (hydrogen in itself may be a free radical neutralizer) and (2) high pressure, an intensifier of whatever other effects are expected. It is not yet clear from these results whether the observations are permanent, whether there are as yet undetected harmful effects derived from the hydrogen environment. We are caught in a cruel bind in that oxygen which supports our lives is also a toxic substance by virtue of its ability to produce the free radicals so damaging to our living processes. The margin of safety is so narrow. Some bacteria are unable to survive in the presence of any oxygen. Observed similarities between the lethality of oxygen and ionizing radiation led in 1954 to the theory that free radicals were the root cause of oxygen toxicity.

A healthy person does not survive with an internal body temperature beyond the normal range of 36°C to 38°C although during strenuous work and in febrile disease we may tolerate for short periods temperature as high as 40°C to 41°C. Denaturation (breakdown) of vital cellular protein occurs so rapidly at 44°C to 46°C that pain is evoked and tissue death comes after a few hours. We may also be able to tolerate hypothermia (low temperatures) for brief durations with body temperatures as low as 27°C to 29°C. In a comfortable, thermally neutral environment, women demonstrate lower heat production than do men. In uncomfortably warm surroundings women are able to dilate their blood vessels more rapidly than men and therefore had less sweating. In cold exposures, women indicated much greater discomfort, perhaps due to their lower metabolic rates. Old people become increasingly erratic in their responses to their environment. Given an upset in temperature, diet, physical activity or emotional stress, the powers of adjustment of the old are weaker than for the young systems.

A relationship between longevity and metabolic rate was first suggested by Rubner and Pearl: the total basal metabolic expenditure from maturity to death in several animal species was about 200 kilocalories per gram of body weight. Fruit flies at elevated environmental temperatures demonstrated shorter longevity than control groups. Or is it the product of longevity potential and heart rate which is the true invariant, regardless of ambient conditions or sex of the species? A few experiments show the effects of exercise on lifespan; for rats on revolving drums, those that exercised lived longer than the controls. This seemed true for young rats but not for older ones. For many animals, multiplying the averaged daily BMR by the maximum longevity potential is a constant, running to 200,000 kilocalories for each gram of body weight. As the size of the species shrinks the BMR per unit weight becomes inordinately large, suggesting that perhaps there is a lower limit to multi cellular mammalian dimensions beyond which the BMR is impossibly large (the higher the metabolic rate, the shorter the lifespan).

Divide the metabolic rate by body temperature. This ratio is a general measure of our entropy production. We live — and die — by descending from a highly ordered system to an increasingly disordered state, towards the ultimate disorder, death. Entropy increases as disorder increases. At a given age — comparing two persons — the one with a higher rate of entropy production is living at a faster pace and less efficiently and will die sooner (if we are allowed to slide towards senile death). As we age, the rate of entropy production is curtailed, but nevertheless continues relentlessly. Thus entropy production may be the tocsin to mark our life's progress: rapid when we are young: slow when we are old.

Our life's processes run their course with some sort of measure of order. The second law of thermodynamics states that every natural phenomenon proceeding of its own volition leads finally to an increase in disorder. Isolated systems tend towards equilibrium with their surroundings where they achieve maximum disorder and maximum entropy. But we need to recognize that we are dealing with open systems where energy and matter can invade our system. It is the summed disorder or entropy produced within us, on and around us, which is always increasing. So if the birth of a baby seems to produce a highly organized (more ordered) state, or the painting of the Mona Lisa is formed from bits and dabs of paints, we must not lose sight of the rest of the universe which becomes a little more disorganized as a result. And the total entropy of the circumscribed system and that beyond, adds up to a higher entropy level than before.

When we melt a solid, its entropy increases by the heat added to melt it, divided by the melting temperature. The material, in being transformed from an ordered, firm structure to a more uncertain fluid has increased its disorder content. The disorder is that of the motion of atoms being mixed at random instead of being neatly separated in the solid. We recognize the natural tendency of things to approach a chaotic state unless we impose outside forces. Every closed physical system changes as its entropy increases, from a less probable to a more probable condition.

Information may be considered a measure of an event's lack of predictability or the dearth of knowledge. It is identical with the number of decisions which are required to explain a phenomenon, or to describe it, or to establish it. Such information increases with the range of the numbers (of symbols or possibilities) and with the disorder or entropy. Order may be formulated from the content of governing laws, multiplied by the number of instances when the laws apply. Life requires a flow of energy from a source to a sink and work must be performed continuously, as we tend always towards the equilibrium of death. Order could correspond to the tension between the storing of energy and its decay as we drift towards the most random possible configuration. When we bring a cold object in contact with a hot one, heat is exchanged so that eventually both bodies acquire the same temperature. The system has becomes thermally homogeneous. The reverse process, however, is never observed in nature. Thus the direction of a natural process is unique. A drop of ink in water will spread until finally the color is distributed uniformly. The spontaneous assembly of ink from its solution to an ink drop is never observed. Consider a moving car whose engine is suddenly shut off. Eventually the motion ceases as the kinetic (motion) energy is abased by frictional forces, converting the kinetic energy into heat (warming the wheels, engine parts, etc.) The organized, ordered energy has been transformed into a disordered form. Entropy, the measurer of the degree of disorder can chart the course of these changes as its value increases. On the other hand we can manipulate a system from beyond its borders, and improve its degree of order-and hence lower its entropy. For example, water vapor in the form of steam at high temperatures is composed of freely moving molecules, seemingly in random motion. Should we lower the temperature, eventually a liquid drop of water will be formed in which the molecules now possess some sort of regular distance between themselves. Their motions are much more circumscribed. At still lesser temperatures, at the freezing point, liquid water becomes ice crystals in which the molecules are arranged in a fixed order. The same molecules are involved in each phase change but we observed some obvious changes and we say the entropy of the ice crystals is less than that of the steam. Order has been imposed. The cost of doing this is derived from the heat exchanged and the work required to move the heat. The entropy content of the outside world in which the intervening forces were at work has been raised, this increment more than balancing the loss of entropy of the steam to water to ice transformation. Thus the entropy of the universe, system and surroundings, has been increased and there is a bit more disorder abounding. Biological processes do not achieve order by the lowering of temperatures but by maintaining a flux of energy and matter. Energy is fed into the system in the form of chemical energy (food) which eventually results in ordered phenomena (growth, protein synthesis, locomotion, etc.). Out of chaotic states we are able to marshal the forces for self-organization.

The property of self-organization is a fundamental feature of living systems. It is reflected in evolution as well as in the development of limbs and other functional appendages. But in getting from here to there the self-organizing systems, particularly living organisms, don't necessarily travel in direct, linear steps, but rather in oscillatory fits and starts. Biological order results from chemical reactions and mass and heat transfer, driving us away from steady, unchanging equilibrium. It is clearly recognized that biological systems are irreversibly organized (they seem to grow and evolve) which is one of the most striking and intriguing aspects of natural phenomena: complex systems, involving large numbers of strongly interacting elements, can form and maintain patterns of order. From the most elemental levels of chemical reactions to the macroscopic development of multi-cellular beings such as we are, or even for societal systems, concepts such as regulation, information and communications play prominent roles. We are in reality alluding to the thermodynamics of irreversible processes, pioneered by Ilya Prigogine who proposed that the transition from an ordered, biological or chemical state to another ordered state requires a critical distance from equilibrium. As we stray further and further from our equilibrium, there is a point reached when our stable state becomes unstable and we then evolve towards a new regime, a new ordered state; the transition is governed by stability criteria based on Excess Entropy and Excess Entropy Production. Systems with the potential for making such transitions tend to have complex relationships describing their behavior: enzyme behavior in our bodies; chlorophyll-induced plant growth; some fluid motions; cell membranes. Fluctuations are an essential aspect of evolution in living systems driven from equilibrium; eventually the decisive fluctuation is produced, the chaotic evolutionary phase is ended and a new stable state is attained. Under these non-equilibrium circumstances, entropy production arises from the internal, irreversible processes (this always tends to increase) and from the entropy exchange with the surroundings (which may be positive or negative). An entropy flow from the surroundings of such magnitude as to overwhelm the internal, ever positive entropy is the fundamental thermodynamic prerequisite for self organization. Living systems continuously exchanging energy and matter beget the instabilities which lead to self-organization. Cells can live only if they are fed their chemical diets which enter by diffusion through their membranes. Within the cells, the mitochondria require oxygen, carbon dioxide or light for the biochemical reactions. The cells are open systems, subjected to various transport processed across their membranes, displacing the metabolic pathways from their equilibrium. Which brings us back to information and its increase or decrease as we roam from one state to another. You will recall that information was a uniquely defined measure of probability, a specification of how many yes-no decisions one needs in order to achieve a definite result within a given number of alternatives. Hence information relates to complexity. In everyday speech, the concept of information is associated with the content and meaning of an item of news. To inform means to supply someone with knowledge or information. Can the meaning of an item of news be made objective and absolute? Consider the deciphering of a piece of news: the news specialist decodes it from the properties of the language, that is, from the frequency of use of certain linguistic symbols, from the probability of their sequence, from the word lengths, the rules of syntax, etc. Thus it is possible to establish objectively, some predictions of the sense of the news, a meaning to it. What happens in the brain of the news receiver is subjective, dependent on previous history. The system, news and receiver, are an inseparable unit. Information is dependent on certain initial and boundary conditions and is self-organizational in that the brain of the news receiver provides the forces to form it into a coherent pattern of recognition. Then is the self-organization of information an accidental, historically unique confluence of events, or is there some recondite regularity at work? Consider the snowflake, a structure of exceedingly diverse detail containing on average about 10^{18} water molecules $(10,000,000,000,000,000,000)$. How are we to code a machine to describe the orientation of all these molecules? And this is but one aspect of the problem which confronts us as we attempt to uncover the genetic coding of one of the simplest of living entities, the genetic material of a coli bacterium, transmitted from generation to generation by the giant DNA molecule (represented by about 4 million symbols in a linear chain, each symbol using elements of a 4-letter alphabet). Such sentences would have the dimensions of a book of 1,000 printed pages; the sequences determining the unique macroscopic characteristics of this bacterium, its ability to metabolize certain substances for energy, self-maintenance and reproduction. There are $10^{2,000,000}$ alternative sequences (10 followed by two million zeroes), a complexity so great that we have practically no chance of achieving by coincidence the correct sequence. This property, to select the proper code and information on the microscopic level we call life. It would seem the life process must fulfill these conditions: (1) the system must possess a metabolism which can build each individual species from energy-rich matter and eliminate energy-deficient products. Metabolism is thus a conversion of free energy, a continuous compensation for the entropy and disorder generated by the irreversible chemical reactions. In this way, the system is blocked from tending inexorably towards an equilibrium state; (2) the system must be self-reproducing. Mutations which may appear because of miscopying can be a source of new information; the system learns from its mistakes.

There has always been a strong connection between living systems and thermodynamics, particularly in the work of Rubner who demonstrated the validity of energy balances in biology and in his study of the metabolism of microorganisms. We recognize today that living matter devolves according to the thermodynamics of irreversible processes, death being the end result of living. We are entropy (disorder) generators; the second law of thermodynamics says, among other things, that life is accompanied by dissipation (heat production). Our living systems are open systems; we add food and air (mass and energy) to our bodies through our boundaries (mouth and skin). There is entropy exchange in breathing as well as that which is generated inside of us. The chemistry of life is irreversible (we can never be the same as before), we proceed with increasing loss and disorder and entropy content. We can measure this entropy production by determining the heat evolved: this dissipation heat, arising from the chemical reactions in us, develops from the degradation of the energy contained in the food we consume. Metabolic reactions, the work of diffusion, osmosis, electrical and mechanical exertions are accompanied by dissipation heat production. One gram of human body weight releases 10,000 times more heat than 1 gram of sun which gives some indication of the intensity of the life process. A running man releases the same amount of heat per pound weight as does a big ocean liner, the tiny fruit fly while in flight generates heat equivalent to a car riding at the highest speeds, when calculated on an equivalent weight basis; a bacterium is in the same class as a jet airplane.

Thermodynamics and information theory and the extent of order in living systems: they represent a connection that may now be obvious. The living system, as an open system, may approach not only equilibrium, but also the stationary state, the difference being that in the latter, the inherent processes occurring across boundaries continue, but at a constant rate whereas in equilibrium, things cease to happen. For example, if I have a tank half-filled with water and allow water to enter at a rate equal to that which is leaving, the tank has achieved a stationary state. No changes are observed in the water level, even though water comes in and exits. On the other hand, should the tank be filled to capacity with hot water and sealed at inlet and outlet and allowed to sit for a long time in an air conditioned room of fixed temperature, we would find that eventually the tank would lose its excess heat and assume the temperature of the room. Thereafter no further changes in tank temperature can be observed and we say the tank has come to equilibrium with its surroundings. In the stationary state entropy production becomes constant and minimal, that is, it is the lowest it can be under the circumstances. In other words, the creation of disorder, while continuing, is at its low ebb. In our approach to senile death, the living system is at low ebb; our rate of living tends towards a minimum, which may be zero or some critical level below which we lack the energy to maintain the life force.

Living organisms, open and exchanging mass and energy, evolve from life to death by dying a little every day. If you believe the wear and tear (rate of living) theory of aging, we may have within us certain total entropy potential. Prigogine and Wiame, and then Prigogine alone, applied the principles of the thermodynamics of irreversible processes to the phenomena of development, growth and aging; they assumed living is a process of continuous approach to the final equilibrium state, accompanied by a steady decline in entropy production (or vitality or rate of living). We are born, develop, grow and age in a continuous manner, the benchmarks being our rate of entropy or disorder accumulation, which can be related to our heat generation and respiration.

As we approach senile death in old age, we become progressively weaker and helpless before stresses, unable to overcome them; they now become killing stresses and so we die, not of monumental upsets but because of an inability to restore our former state. Is there a critical level of heat or entropy production below which it is impossible for the living system to support life (our vitality is simply insufficient to maintain the tension of life?).

Fish continue to grow during the entire lifespan and therefore growth and aging are difficult to differentiate. Mammals and birds are more interesting in this respect since here growth proceeds only in the first third or half of the lifespan. Insects also reach terminal sizes; human growth stops approximately at the age of 20 to 25 years; after this age the BMR diminishes at 3 to 7 percent per decade, even in the last stages of aging. But do we achieve that final stationary state where heat production actually reaches a minimum? Zotin and Zotina say we ineluctably move towards such a final stationary state, a condition synonymous with natural death and suggest we can study embryo development, growth and aging through entropy analysis: the decrease in embryo respiration during its development may relate to the aging process. The mitochondria within the nucleus of our cells, the energizers, increase in concentration in the rat brain between 3 to 33 days after birth, concomitantly with ascending respiration in the brain tissue. And then the brain experiences a loss of mitochondria. In the human liver, we lose them quickly after age 60 and of course simultaneously see a fall in oxygen respiration. Each new organism begins its development from a maximum level of heat or entropy production. This ability to start from zero, achieve a maximum and then decline is of utmost interest not only in the study of normal aging, but also as we explore regeneration of organs and tissue, wound healing and malignant growth. Do we collect the energy or entropy credits, and then, sufficiently activated, enter the pathway of development and growth? The amputation of the posterior part of an earthworm at segment 60 is followed by an increase in oxygen respiration, attributed to the wound infliction. Soon respiration diminishes, followed by a new wave of respiration stimulation which is related to the regeneration of the earthworm segments. This reflexive respiration increase is found not only in the earthworm, but also in mammals and other animals during regeneration of the liver. Maximum metabolic rates are attained somewhere in the middle of the regeneration stage and then gradually return to normal. Do we need an energy expenditure of unusual intensity to drive the cells towards repair or regeneration? Some research suggests this, the work having been done on axolotls of various ages, which showed approximately the same critical level of oxygen respiration. In some dissembled way cancer cells represent rejuvenated cells, with high rates of metabolic heat (entropy) production, but their rejuvenation doesn't lead to resumption of the normal process of aging; cancer cells attain a high metabolic rate and maintain it. With malignancy there may be an augmented concentration of mitochondria in the cells. For living systems at rest, with the chemical reactions proceeding relatively slowly, all of the internal heat generation evolved become equivalent to entropy production and is measured by the heat lost to the environment. The living system with age passes from a less probable state to a more probable one: the most probable state being death. In the course of development, growth and aging, each event is accompanied by its characteristic heat evolution. Thus the continuous recording of heat evolution ought to detect fluctuations of the vital functions in sub cellular biochemical systems as well as in isolated organs or in the whole organism. These measurements have been made in the past. Drinking water can provoke a sweating response and heat generation. Patients with different kinds of anemia showed elevated rates of heat production. If the heat cannot escape from the environment which has captured it, as in microbial accumulations in hay, wool, manure, and alcoholic fermentation, spontaneous combustion and fires can result. Temperatures as high as 265°C can be found in hay. If there is no heat regulation, living systems will die. For humans, special mechanism have been developed to dissipate the heat; through blood circulation for example. (We maintain a constant temperature independent of the varying environmental temperatures.) Yet in a beehive, a constant temperature is maintained summer and winter by the collective contributions of all the bees in the hive, whose individual heat evolution is variable but controlled to produce the constant ambience. In microorganisms, a large part of the heat generations is due to the cell division process, whereas for higher animals such as humans, the heat originates mainly from muscular contractions and from glandular activities controlled by the nervous system. Bacteria exhibit higher heat production (and entropy generation) than we do, on a unit weight basis. Somehow we have learned to moderate our rate of living, perhaps by controlling our body temperature.

Though the aging process has been studied extensively, no one theory has been advanced to answer all the questions raised. Since entropy seems to be one of the premier variables in nature, which may at times parallel the direction and irreversibility of time, it appeals to many as a powerful tool in longevity analysis. When the question of the appropriateness of entropy calculations for biological systems was submitted to an international conference at the College de France in 1938, it caused much acrimonious debate and no agreement could be reached. The difficulty focused on the fact that living systems must exchange matter and energy with the environment in order to survive and hence are open systems, not in equilibrium with the environment. Prigogine and Wiame published an extended form of the second law of thermodynamics which applied not only to isolated (closed) systems but also to open (and possibly living) systems. Subsequently Prigogine divided entropy generation of an open system into two parts: the entropy flow due to exchanges with the surroundings (food and inspired and expired air for a living person) and the internal entropy production of irreversible (chemical, metabolic) processes. Internal entropy production is always zero (if we could live at an infinitesimally slow rate) or greater than zero (as we live normally and age irreversibly), but the external entropy exchange can have any sign, depending on what comes in and goes out of our bodies. We are open and maintain ourselves by the transfer of food and air (energy and matter) with the environmental and by continuous chemical synthesis. The metabolic rates of a living organism of necessity must not be too slow; hence they contain a large number of irreversible processes (chemical reactions) which contribute to the internal entropy production and its concomitant effect, the heat evolved. Maximum entropy (disorder at its highest level) is perhaps equivalent to senile death (not caused by accidents or any unusual intervention.) It may be that death is lurking when our lifetime accumulation of entropy approaches a predetermined, fixed allocation for our species. Or when our rate of production dips to some lower critical minimum level where we are faced with a loss of vitality and are weakened sufficiently so that any one of a number of life's minor stresses (minor when we are younger) becomes a killing stress. The onset of a cold in a very old person frequently leads to pneumonia and complications which undermines the general health and vigor of the old person to the point where he or she finally succumbs.

As discussed previously, for the human in a basal state, essentially all the energy output from the catabolism of food in the body appears as heat and the rate of internal production of entropy related to its metabolism surpasses by far that connected with other entropy flows. So now we know that the internal entropy production can be calculated from the heat of the chemical reactions in our bodies (divided by body temperature) and that this heat can be obtained from the BMR. The external entropy exchange is obtained from knowledge of the amount and composition of the inhaled and exhaled air we breathe. The sum is our rate of entropy production. From these ideas, we find the expected lifespan for males in general is around 84 years; for females the corresponding figure is 96 years. (Based on Metropolitan Life Insurance data these numbers become 103 and 109 years respectively for males and females.) If there is such a thing as a critical BMR value, we would expect this at senile death and we find this critical BMR value to be 0.84 kilocalories per kilogram weight per hour for males (0.83 according to Calloway). This translates into a critical entropy production in the vicinity of death of 0.00269 kilocalories per kilogram weight per hour per degree Kelvin temperature for men; the corresponding value for women was 0.00260. The wear and tear (rate of living) theory of aging allows that we have within us a programmed amount of life "substance" which is consumed in proportion to the duration and manner of living. When we deplete it or if the residual drops to some critical level, or the rate at which we use the elemental material diminishes to some sensitive mark, we are vulnerable to death. Hershey and Wang have pooled the BMR data of the last fifty years in order to identify average, typical entropy production curves. Where the curve levels off after its steady decline with age, at age 84 years for men, death is imminent. Prigogine and Zotin independently proposed this as the criterion of impending death: the principle of minimum entropy production. From the Hershey and Wang composite curves; the total lifetime accumulation of entropy for men and women is 2,395 and 2,551 respectively, in units of kilocalories per kilogram per degree Kelvin. This then is our potential. We can calculate your specific cumulative entropy production from birth up to a given chronological age and compare this with the average cumulative figure for a human subject. If your number is higher than average, you are living at a faster pace and are older than average since you are depleting the lifetime potential at an accelerated rate. How long you live depends on when you reach the neighborhood of 2,395 (men) or 2,551 (women).

PART IV

Entropy Theory of Aging Systems: Humans

CHAPTER 7

Entropy and the Life Process

The Second Law of Thermodynamics

The second law of thermodynamics speaks to the concept of Infinity. The second law says that all things, if left to their own volition, will eventually disintegrate. They will trace out the birth, growth, maturity, senescence, and death sequence. Put a wall around a house and don't let air, water, food, electricity, energy, matter, or people in or out, and eventually everything inside will die and go to... Infinity. The old Soviet Union disintegrated because in maturity they placed a virtual wall around themselves, and let the infrastructure of the empire become decrepit. It collapsed, "killing" the Soviet Union. We speak of a closed or isolated system as one that is impervious to the transport of mass and energy and information through the "membrane" surrounding it. We also say, and can calculate, that this isolated dying system has attained maximum entropy, a prescription for disaster, and has reached the final equilibrium state (death) where all driving forces for change are gone.

Entropy, Time, Information, Infinity

Entropy can tell us something about time's arrow, about the direction of time. Measure my entropy now and ten years later. I will look older in ten years and my entropy will be higher than it is now. Entropy maps the degree of order and disorder: higher entropy indicates more disorder, which means I can calculate my entropy ten years hence and confirm that its higher value shows time has passed, from my present to my older future. I can tell which came first, the green leaf of spring or the red of fall, by calculating the green leaf entropy and the red leaf entropy. The red's will be greater, telling me from the second law of thermodynamics, that this leaf has evolved from lower to higher entropy, from spring to fall. I know spring came first because spring green has less entropy than fall red.

I can take an ordinary book, with its letters organized into words which are strung in sentences on the pages which are bound together in a book, whose volume is contained in a library of books. I can rip these pages from the book, and with my scissors cut each page into individual words and each word into individual letters. I can pile these individual letters on the floor. And what have I? I have much more than just a pile of junk letters, the remnants of the organized assembly of letters which became the words and the sentences for the plot of the story of the book. For what that pile of individual letters contains is a near Infinity of information, the potential information for writing not only the original book, but many, many other books. Simply by rearranging the letters piled on the floor. That pile of letters is like the infinite potential information of Infinity, from which is obtained the organized and stored information of our book. The book and the library represent the finite world of our limited intelligence.

Infinity and Entropy

If Infinity is entered through the porthole of maximum entropy, then how did we become our universe? How did we separate from Infinity?

Scientists believe our world, our universe, began with a big bang, but the bigger question is, where did the big bang come from. Within Infinity (which contains an Infinity of potential information) it is possible for pockets of low entropy and order and stored information to form spontaneously. It was one of these vacuoles or islands of stored information, which became our universe, our world, our galaxy, our Milky Way. We are ephemeral states, existing so long as our entropy content stays below maximum entropy. Inevitably we achieve maximum entropy and "see" the entry port, back into Infinity. We will dissolve into the ether of Infinity. Within the context of Infinity and its eternity, our universe and our existence on earth are transitory and will be relatively short-lived. If we could calculate what constitutes maximum entropy for us on earth and within our universe, and if we calculate what is presently our entropy content, this could be very meaningful. The difference between maximum entropy (where we are heading) and entropy (where we are presently) is the distance, in entropy terms, to the end of our world, when we will return to

Infinity. This entropic distance is called Excess Entropy. The speed at which we are approaching maximum entropy is called Excess Entropy Production.

Excess Entropy and Excess Entropy Production

Entropy is a measure of the disorder in a system.

Entropy increases as differences or tensions within the system are dissipated.

Entropy increases as the size of the system is increased.

Entropy tends towards a maximum in the vicinity of death, as control is lost.

Excess Entropy (EE) is a measure of the entropy distance from disaster. EE approaches zero as we approach disaster or chaos or death.

Excess Entropy Production (EEP) is the rate of change with time of Excess Entropy (EE). EEP diminishes with age and nears zero in the vicinity of death. For the aging system, a diminishing EEP can signify stagnation and a general decline in vitality.

The driving force for change, the motivating factor which drives us beyond one non-equilibrium state to the next may be Excess Entropy. It expresses the tension of life, the distance (in entropy terms) from equilibrium (death, disaster, chaos). Excess Entropy Production (EEP) measures the speed of approach to equilibrium.

Entropy and Lifespan

Entropy is related to energy, or more specifically to the quality of this energy. This means the poor quality energy coming out of the tail pipe of our cars will be discarded. We think of the exhaust gases from our cars as useless and not capable of doing work for us. But the same amount of energy coming from the burning of gases on our stove is used to heat food placed in pots and sitting on top of this energy. The difference between the car exhaust energy and the energy on our stove is the temperature of this energy. Energy at or near room temperature isn't very useful, but energy at high temperature is deemed to be very valuable. So we need to distinguish between the energy of the auto exhaust and the energy of the stove. We can do this by dividing this heat energy, Q, by its temperature, T. The ratio, Q/T, characterizes the quality of the energy, its potential for doing work, and is related to entropy. Entropy in the beginning was defined as the quality of energy which makes it useful and available to do work, to cause changes. Eventually, we noticed that chemical reactions and other natural processes proceeded from reactants (with low Q/T ratios) to products (with high Q/T ratios). This was the direction things went, if there was no outside intervention: from low to high Q/T ratios, or in other words, from low to high entropy. It is true for chemical reactions and in the factory as well, and even in our own bodies.

We age, and with time, we generate entropy. This increasing entropy finally approaches maximum entropy, a condition of ultimate disorder, and death. Not only can we calculate our entropy, S, at this moment, but we can also calculate maximum entropy (S_{max}), towards which we are attracted. Thus we can know ($S-S_{max}$), which is Excess Entropy (EE), our distance from disaster (death). We can compute a new age scale based on this Excess Entropy (EE), and the velocity with which we are approaching maximum entropy, Excess Entropy Production (EEP).

Natural (senile) death will occur when we achieve maximum entropy, S_{max}, which means Excess Entropy (EE) goes to zero. Also, we know that in the vicinity of death, we achieve a sort of equilibrium state, where all driving forces for change have disappeared, which means they have become zero. Thus death, the ultimate disorder, occurs when we achieve maximum entropy, S_{max}, and when the Excess Entropy (EE) and the Excess Entropy Production (EEP) go to zero, when all driving forces for change have disappeared. Our calculations indicate senile or natural death should, on average, occur at about 85 years for males and about 100 years for females. Middle age has now retreated into the late sixties and early seventies, where old age begins.

Even so-called inanimate systems such as corporations and countries and civilizations are subject to the entropy laws of life. We can compute maximum entropy for a corporation by defining disaster as when all the units of the organization are out of control. The organization in this state is at maximum entropy and in the vicinity of death. In a manner similar to what we showed for living systems, for this so-called inanimate organization, we can also calculate the ($S-S_{max}$), the distance from disaster or death, Excess Entropy (EE). We can also calculate the speed of approach to disaster, Excess Entropy Production (EEP). Since inanimate organizations can be rejuvenated, they need not die. Instead they may evolve into more viable forms of life with lower entropy and greater EE values. Mutations in nature, in living systems, may be a way of forcing a greater distance between entropy and maximum entropy, and giving new life (rejuvenation) to a species.

Order

Order (and disorder) are linked by the second law of thermodynamics. We can intertwine physical and mechanical order and disorder with entropy. We find geometrical order by examining crystals and molecules. John Cage showed us that less is more (sometimes) with minimalist musical exercises which control us (order us). The unifier of knowledge is entropy. And are we not controlled by the limitations imposed by maximum disorder (maximum entropy).

In order there is also information. Transmitting information induces order. Information gives form, and form needs structure. Entropy measures the dissipation of energy, of information, of truth, of life, of being. Entropy controls transitions, from A to B, from chemical reactants to chemical products, from life to death, from death to God and Infinity. Jackson Pollack, an abstract expressionist painter, sprinkled and splashed pigment on his canvases and found order through fluctuations. He found the path, from a stable state, to a disorderly creative process, into a new stable state. At this bifurcation point of instability, it could have ended badly and absurdly — one path was disaster — but he found the other path, to exciting new forms of expression in his paintings.

We can be content to stay with the status quo, to stay comfortable, at or near equilibrium, where no significant changes can occur. In this equilibrium state, small driving forces for change can be met with resistive forces sufficient to return us to the equilibrium, unchanging state. No new ideas are developed here. Nothing new is learned or done or dreamed. Yet there is always, with time, an accumulation of tensions pressing for change, this amid the ambience of internal deterioration predicted by the second law of thermodynamics. If a driving force for change develops, and is large enough to push us far from equilibrium, then a return to the comfortable equilibrium state becomes virtually impossible, and we are faced with the new challenges and choices. At this moment of bifurcation, we can elect to travel the unknown path with its promise of something new, or we may inadvertently go down the path to disaster and maximum entropy. In any case, we may really have no choice. Change is the norm. Otherwise, we will become increasingly disordered and disintegrate from within, a victim of the workings of the second law of thermodynamics.

We relieve tension by simplification and change, by inducing order. Basically, this is the law of evolution espoused by Spencer. In music, the simple chaotic chanting of primitive tribes yields modern complex musical forms. What could be more interesting than an ellipse, incorporating simplicity with constantly changing location, angles and symmetry?

"We encourage others to change, only if we can honor who they are now. We ourselves engage in change only as we discover that we might be more of who we are by becoming something different. We organize always to affirm and enrich our identity. Too often, highly structured organizations destroy our desires. They insist on their own imperatives. They forget that we are self-organizing. Whether we are beginning a relationship, a team, a community organizing a big project, or a global corporation, we need together to be asking what we are trying to be? What's possible now? How can the world be different because of us? In too rigid organizations we focus on techniques for policing or enticing one another into behaviors and roles. Instead we should support our ability to self-organize. Order through freedom. The world is inherently disordered but we seek organization. Life is attracted to order." (Wheatly)

Driving Forces for Change

The laws of life, of our world, of our universe, incorporate science, order, God, Infinity and entropy. Whether we examine DNA, the big bang theory, or evolution, the focal points are the systems which are self-organizing. The origins of life, the theories of life, cell behavior and the crystallization of life's matter revolve around self-organization, entropy, Infinity, and God. These are catalysts which provide the energy for change. Spontaneous regeneration and growth affect complexity and size. We live on the edge of chaos, and flow towards maximum entropy, while maintaining the ability to organize ourselves and our world in some spontaneous ways, driven by forces not always within our realm of knowledge.

Cells which become too large to be viable divide into smaller cells and begin the growth process again. Corporations do the same. So do the stars. We speak of dissipation systems, such as living people, those that proceed with increasing entropy (and disorder). We speak of entropy driving forces, between the entropy content we have now, compared to our maximum entropy content of maximum disorder (death). We speak of these being nonlinear processes, of being propelled by the entropy driving forces for change. We speak of stability through fluctuations, as we march through these changes, allowed to find new places to end up, without big upsets or explosions. We speak of arriving at the fork in the road, the bifurcation point, where there are new alternatives available, the so-called thermodynamic branches. Choose one, and we may find a nice smooth transition to a new condition, a new place to be. Choose the other branch, and perhaps chaos and disaster lurk. But one thing we always know, the second law of thermodynamics dictates that change is the norm. That you'd better be ready to change, for the plan to remain permanently in the status quo, in the equilibrium state, is a prescription for disaster. Never change, and we will predictably disintegrate from within, the rot will begin and continue until there is no core. We know that we should never be isolated, or insulated, against the forces for change. We know we must remain open systems, open to the exchange of energy, matter, and information from those beyond our borders. We know that we and our world proceed through various stages, all the while generating more and more entropy. When we land in a new state, we settle in, dust ourselves off, rearrange our "feathers" (lower our entropy production) and stay for a while, awaiting the next push to another stationary, non-equilibrium state. We reside in each new state guided by the stricture of minimum entropy production, which is really what we are doing when we settle in, dust ourselves off, and rearrange things. We know the driving force for change is Excess Entropy (EE), the difference between our current entropy condition and maximum entropy. The driving force for change is also Excess Entropy Production (EEP), the speed of approach to maximum entropy.

Put a shallow pan of water on the stove, at a low heat, and watch from above. Under the right conditions, you will see the spontaneous formation of a structure. The water surface seems to be self-organized into tiny cells. It happens spontaneously, surprisingly, driven by those entropy forces. Or you can take a cylinder of water with molecules A (heavier) and B (lighter) dissolved in the water. The A and B molecules are uniformly distributed as long as the temperature is the same throughout the cylinder. Now heat the top of the cylinder while cooling the bottom, and wait. With our measuring instruments we can soon detect that we have created a concentration difference, driven by the heat flowing from top to bottom. The heavier A molecule will concentrate at the bottom while the lighter B molecules cluster at the top. As long as we maintain the top hot and the bottom cold, the concentration gradient will remain. The new condition is a different state than the original one of uniform concentration of A and B. It will disappear as soon as we remove the driving force of heat flow (temperature gradient).

I can do anything if I do it slowly enough. I can do anything if I take a very long time to do it. I can do anything if the pressure or force I apply is very small, and just large enough to overcome the resistance to change. It is more efficient to boil water in a pot on my stove by heating the water with a temperature always only slightly higher than the rising water temperature. If the water is 80°F, the heat applied might be 80.5°F, if the water is 100°F the heat is 100.5°F. From the second law of thermodynamics, which up to now has addressed the issue of entropy, we can also use it to mine the concept of efficiency. From the second law of thermodynamics we can expect that the most efficient processes are those driven by forces which are always just infinitesimally larger than the resistance in the way of change. So when we are heating the water on the stove, if the water temperature is 100°F, it would be theoretically most efficient to apply heat of 100.0001°F. We have here a driving force (the difference in temperature, between the water and the heat) so small that it tends towards zero. The process is most efficient, but of course it would take a very long time to get the water temperature to rise to, for example, its boiling point of 212°F. Theoretically, the efficiency of this heating process approaches 100% when the driving force for change is so small that we say it approaches zero. We would say, in this case, that the heating process is being done reversibly, and at 100% efficiency because it is so slow, taking an infinite amount of time. If done reversibly, this heating process could become a reversible cooling process, if the cooling surface is now at an infinitesimally smaller temperature than the water. For example, if the water is 100°F, while the cooling surface is 99.999°F. When the water has cooled to 80°F, the cooling surface might be 79.999°F. Reversible means we can go either way, up or down, by simply making the temperature driving force for heating or cooling be infinitesimal. The real actions in our world proceed not reversibly by very, very small driving forces, but by finite, sizeable driving forces, and therefore are not theoretically reversible. We, at present, don't know how to reverse the aging process. Leaves turn green to yellow to brown in the fall, and drop from trees. We don't know, at this moment, how to reverse this, and get the brown leaves on the ground to reattach themselves to the tree limbs and turn green again. These irreversible processes, like aging, go from low entropy to high entropy. In order to move time backwards, we would need to find a supernatural outside force to lower our entropy and reverse aging.

Steady State, Unsteady State

Suppose I have a tank of water. There is water standing in the tank, and water can flow in and exit through the drain. If the same amount of water flows in as goes out through the drain, then the water level in the tank remains the same. There is water coming in and water going out, but to all appearances, nothing is happening. The same amount of water comes in, the same amount of water goes out, the level of water in the tank is unchanged, and we say the tank is at steady state. Nothing is changing with time. Everything is in balance: in minus out equals zero. We say this is a steady or stationary state, where all kinds of things are happening but nothing is changing with time. Now suppose I change slightly the amount of water coming in by adjusting the inlet valve. For a brief period there is more water coming in than there is going out. This causes the water level in the tank to rise. A higher level of water in the tank will force more water to exit through the drain than before. The water level in the tank will continue to rise (causing water to leave through the drain at a faster rate) until the increased water level forces the exiting flow rate to just equal that which is entering. Then the level of the water in the tank will cease to rise and we are in balance again, at a steady state. The tank is in a new condition, a new steady state, where the flows in and out and the water level don't change with time. But this second steady state is different from the first. Evolution proceeds this way, aging proceeds this way, corporate life proceeds this way, as does everything else: from one steady or stationary state to next, to the next, etc.

Contrast this steady state condition with equilibrium. Suppose I have a piece of iron which I've kept for a very long time in my refrigerator. The iron is just about at the temperature of the refrigerator, say 40°F. Now I remove it from the refrigerator, place it in a large room which is maintained at 70°F. The piece of iron in the room will begin to warm. It will acquire heat, the heat transfer being driven by the temperature difference between the room, 70°F, and the iron, 40°F: (70° − 40°). We know the piece of iron will begin to warm, so now the iron is 50°F. The driving force for heat flow is now smaller (70° − 50°), so less heat transfer is occurring. Now the iron is 60°F, and the driving force is (70° − 60°) and still less heat flows into the iron. Eventually the piece of iron gets to 69°F, which means the driving force is (70° − 69°) and a very small amount of heat is flowing into the piece of iron. The iron is now 69.5°F, then 69.99°F, then 69.9999°F and we expect that the heat transfer into the piece of iron is now minuscule, driven by a tiny temperature driving force (70.0000°F − 69.9999°F). The driving force for change is approaching zero, which means the amount of heat flowing into the iron is also approaching zero. How long would it take for the piece of iron to get to exactly 70.0000°F? The answer is, an infinitely long time, since the heat flow tends towards zero as the driving force tends towards zero. If we waited until time went to Infinity, the piece of iron will have achieved 70.0000°F, and we would say the temperature driving force for change is zero. We would say that the piece of iron is in equilibrium with its environment, where at equilibrium, not only are things not changing with time, but also the driving forces for change have disappeared and no further change in the iron will occur unless we alter something (room temperature, for example). No further change will occur unless there is outside intervention.

Steady state and equilibrium are similar in that there are no changes with time, but equilibrium is a stronger condition, since at equilibrium, the driving forces for change and the flows have disappeared. Death is the final equilibrium condition, where flows and forces are zero. We cannot overcome the killing stresses acting on us and we are at maximum entropy, maximum disorder.

Self-Organization

Self-organizing systems evolve through spontaneous increases in organizational control and complexity. A property of self-organizing systems is the ability to learn, that is, self-modification, which is a spontaneous emergence of order, far from equilibrium. The convective flow pattern, Bernard cells, formed by water being heated in a shallow pan, is an example of a self-organizing system. The totality of life on earth behaves in some ways like a living organism, and seems to regulate itself in the short run (constant oxygen content in the atmosphere or the salinity of the oceans). But we know that the present self-organizing earth state is simply one of an Infinity of stationary or steady states possible. It is only a matter of waiting long enough to see the evolution from one stationary state to the next. For example, there can be biological evolution, as well as self-organizing and evolving linguistics, crystal growth, social processes, committees, and task forces-all self-organizing and evolving.

If a salt solution of sodium and chloride ions in water dries out, what remains are salt crystals, sodium and chloride ions now locked in an unalterable structure. Self-organization. Might life have begun so simply, where from some sort of primordial soup there arose some combination of matter which self-organized into something functioning, eventually something capable of reproduction? Self-organization and then evolution into other forms. The formation of structure is a key to self-organization.

But what drives the changes? What forces us to change? Is it a social force or the power of numbers present, or is it an entropy force, directing us to pay attention to order and disorder, and to evolve to new forms far from maximum entropy?

Open Systems and Closed Systems

Our living bodies are open systems. Things are happening beneath our skin: chemical reactions; enzymes diffusing from one location to another; blood flowing; etc. At the same time we are inhaling air (and oxygen) and exhaling air (with less oxygen and more water vapor and carbon dioxide than before). We are also ingesting food and eliminating waste products. Some chemicals in contact with our porous skin may diffuse into our bodies through our skin. Water may effuse from the interior, to the skin and through it, perhaps ending up as sweat. We are what we are as a consequence of what is happening beneath our skin plus what makes its way through our boundaries (skin and membranes). The sum of internal happenings and external transport becomes who we are and what we are to become. Cut off the external flows (oxygen, water, etc.) and inside we suffocate and die very quickly. But these external flows moderate our life's process, slowing the accumulation of entropy and stretching our lives to eighty-five to a hundred years. Open systems allow for the transport of energy, mass and information through our borders. Close the borders and you have rapid internal destruction and rapid accumulation of entropy, as dictated by the second law of thermodynamics. Hadrian's wall in England, the great wall of China, and even the closed minds of the old Soviet empire are examples of the ill-considered attempts to control the borders.

Closed systems do not self-organize, they self-destruct. This means we need to assure that all the things we do, all the systems we work with, are open. We must remain receptive to the exchange of information, energy, and mass between the external environment and the internal working units. We need to pay attention to the size of our system, the geometry of the internal structure, the duration and intensity of the forces for change, the degree of exchange of information within the system, the nature of the external environment, and the quality of the information available. Is the border or membrane surrounding us stable and porous?

Aging and Evolving Systems

If only we could remain the same, you and I, youthful and strong, resilient and hopeful, ever-learning, never weary, always able to rebound to meet the next day. We express the plea for a status quo, a steady state, but only if conditions are perfect or nearly perfect for us. For some, the goal is not to remain unchanged, not to continue repeating in some way that which is our present lot. No, the aim is for evolution, an ascendancy from the present to the future during which we will raise our station in life, improve our health if this is important, make more money, raise a family, build a house. Do whatever is required to make the transition.

Do you ascribe to the steady state or the evolving-state doctrine? Do you prefer to remain close to some sort of equilibrium condition and allow that small perturbations can be handled in a way that returns us to our original state? Do you regard with suspicion those who proclaim there is an entrenched, privileged class and an oppressive bureaucratic structure which must be torn asunder in order to build anew from the ashes of the old? Are there those who see change as not only possible but necessary for the orderly transition — the gradual evolution — from one stable state to another? Of course, there are persons who seek change for all these and other reasons. Each group representing another philosophy exerts its own pressures. The changes sought may be for the better or may be disastrous, depending on the goals and whether we can control the change agents to some extent.

Is the life of a living system, such as a human being, to be seen as a wear-and-tear process as the individual winds down, or is it a gradual evolution from birth through intermediate states to maturity, senescence, and death? Death is inevitable for all living organisms. For structures such as corporations, it is not readily apparent that they age and die, though some will go bankrupt, which is a form of death. Civilizations die in some way. There have been golden ages for Greece and Rome and perhaps even for the United States. Who can say that the forces at work on civilizations are not similar to those which stress a living human system?

To aver that there are various change possibilities for a system is of course to state the obvious. The changes can take a number of forms. For example, place yourself on a closed loop and allow that things will change in such a way as to trace a path always on the loop. There are limits to the deviations and under no circumstances are conditions sufficient to drive you off the loop. In this world the perturbations from equilibrium never get out of control. If the speed on a highway is 55 miles per hour and you are law abiding, you might average that speed yet there will be times when you travel at 50 or 60 miles per hour. But you always strive to return to the speed limit. The earth in its rotation around the sun traces an orbit which is cyclic and predictable, and is an example of this neutrally stable state.

On the other hand, some systems behave in a manner called unstable equilibrium. This could be the case of a poorly designed nuclear reactor, one with improper temperature control. The heat generated by the nuclear reaction stimulates an increasing rate of reaction which yields additional heat which boosts the reaction rate further, developing still more heat, and so on. Or perhaps we are tracing the loss of middle-class persons from the inner cities. With the departure of some, the tax burden increases for those who remain, schools deteriorate, crime increases, and more middle class persons flee to the suburbs. This then continues the exacerbation of the initial problem — only now the level of the difficulty has been raised and more flights occur. Gradually, steadily, conditions escalate and we head toward a seemingly doomed situation unless there is outside intervention or the constituents themselves change the pattern.

But there are still other possibilities for systems to change. Some negotiations, between persons or other complex organizations (such as successful collective bargaining sessions), trace out such as history. There is a narrowing of differences and a graceful evolution toward a final state, a compromise. Such system behavior usually is aided by inherent moderating influences which prevent blowups; that is, they prevent an uncontrolled growth of the stresses or change agents. It is as if the system were being reined in relentlessly by forces outside it. There are other possibilities for system behavior. A spring-wound pendulum clock, starting at rest, will commence to hunt for its final stable cycle. Should the pendulum be given an unusually large starting shove, the system would gradually descend toward the same stable cycle. A spinning top, no matter what initial velocity imposed, will always end at rest. Death as the ultimate end is approached by all of us in different ways, but the preordained fey conclusion is the same.

The dinosaur probably evolved toward a critical size beyond which it was impossible to support the immense body weight. This increased girth could have affected the dinosaur lifespan by causing difficulty in finding sufficient food to sustain life. The cockroach is believed to be an example of an evolutionary process where the cockroach has changed very little during the long history that humans have observed this species. Further cockroach development is seemingly foreclosed by its equilibrium teleological state.

Instead of seeking the security of constancy, proponents of non-equilibrium dynamics say that organizations proceed to and through various plateaux, resting on each level until displaced by the convergent forces of energy, material, and information exchange with the environment. The human body can be seen anew in this light; humans maintain a relatively stable state until an illness changes their lives. A corporation operates under known, relatively constant conditions until it merges with another corporation. And so it goes, these open systems are capable of being driven from one seemingly stable orbit of operations into another. The system attempts to remain viable by switching to a new dynamic regime, yielding, in a sense, order through fluctuations, which is the reverse of the behavior of some systems near equilibrium. The organizations near equilibrium attempt to meet new pressures for change by damping them out and returning to their original conditions after small, temporary deviations. Some might claim that the forces of racial integration were like this, where, for a while, the integrationists' pressures were met by sufficient resistance to continue the original segregated condition. There seemed to be stability in the segregated state but this was specious since the time frame was not sufficiently long to see the trends. Actually, the segregated state was inherently unstable. When the forces for integration became morally irresistible, the system moved to a new nonequilibrium position, which is the present point in history. Society found order through fluctuation or change. In the new state of non-equilibrium, order may increase, and in this state our response to old pressures is met with more degrees of freedom. There are more ways to meet old problems. With the new dynamics of order through fluctuations, the challenge is to delineate the bounds of stability and to attempt to identify the new state in which we are or to which we wish to go. There can be surprises along the way, of course. For example, a system can be driven too far in size and complexity, as some think New York City has been. The high density of the population and the limited system of roads in Manhattan has led to the surprising result that during rush hours it is often quicker to walk in Manhattan than it is to drive the same distance in an automobile.

These dissipative, non-equilibrium states might in some simple cases be described by mathematical equations with appropriate feedback relationships. If the feedback is negative, this feedback or return of knowledge and information to the pressure point is considered a control on the pressure fluctuation. The tendency is toward stability and a subsiding of the system to its original state. Should the feedback be positive, we have a condition where there is reinforcement of the amplitude of the fluctuations. (We add a positive pressure to a positive feedback.) This yields a higher output, increasing the positive feedback value, which when added to the positive pressure gives a still larger output.

Whether considering organizations such as corporations, societal structures such as the welfare system, human living organisms, or the universe, what emerges is a general schema for change, a way of going from one non-equilibrium state (of disorder) to the next non-equilibrium state: order through fluctuations. The change process is accomplished by deviation-amplifying or positive feedback means. This might be called a revitalization process requiring explicit intent by members of society. To get there from here, the system needs to go through the following steps: (1) achieve a non-equilibrium plateau; (2) experience stress; (3) endure cultural distortion; (4) plan for revitalization; and (5) enter a new non-equilibrium plateau. The manager of such a system is the catalyst rather than the designer of an organization. The task of management under non-equilibrium conditions is to stimulate the growth of a network of decision processes. One objective of this network would be to maintain or renew the autonomous unity of the organization. In other words, it becomes imperative for proper functioning of an organization to allow and guarantee the various integral parts of the system their degrees of freedom within the constraints of the planned operating mode.

Social theories have traditionally been geared to structure, not process, and to ideals of equilibrium and structural stability (stable equilibrium). The emphasis has been on steady state and negative feedback (which corrects deviations). It may be that the analysis and planning of social organizations now may need to be brought into consonance with the newer non-equilibrium order through fluctuations approach. Contemporary society is characterized by a rapid dissemination of information. In the past, science and technology were chief agents of positive feedback (deviations from existing norms were amplified) in transforming the human-environment relationships, such as the discovery of the deleterious effects of the pesticide DDT and its subsequent discontinued use. On the other hand, societal philosophy and techniques provided negative feedback to stabilize the relationships (zoning regulations within cities, for example). But the deviation-amplifying process can increase differentiation, develop structure, and generate complexity. This may work to enrich society and allow us to move on to something better than we had previously. Much of the unpredictability of history is perhaps attributable to deviation-amplifying causes. The result is either a runaway situation or evolution.

A New Vocabulary

- Dissipative systems
- Order through fluctuations

- Self-organizing processes
- Equilibrium
- Non-equilibrium stationary states
- The second law of thermodynamics
- Open systems
- Entropy
- Positive feedback
- Evolution
- Bifurcations
- Maximum entropy
- Minimum entropy production
- Excess Entropy
- Excess Entropy Production
- God
- Infinity

The Meaning of the New Vocabulary

We are born as dissipative systems, entropy generating, growing, differentiating, and changing. There is order as we change, order through fluctuations, as we grow into childhood, grow taller and stronger. We are being driven to change, self-organizing, by forces hormonal. We are programmed by genes conditioned through the ages, to be formed a certain way. To evolve. Ontogeny recapitulates phylogeny. These forces drive us far from equilibrium, far enough away from equilibrium so that we cannot return to where we were. We are driven beyond the point of return, beyond the status quo. Far from equilibrium, at this bifurcation point, most of us find the path towards the next stage of our development — from childhood to young adulthood and maturity. The other bifurcation path leads to disaster, ill-functioning bodies and perhaps early death. At each stage in our development, at each stationary state we hunker down, collect ourselves, come to grips with what we have become, minimize our entropy production, and get into a steady, comfortable state. We are open systems awaiting the next push towards the next level of development. This pressure comes from a positive feedback mechanism, which creates an added impetus for change. This positive feedback driving force, which may be hormonal, when added to the ambient chemical reactions already within us and driving us, becomes the irresistible force which moves us again beyond the comfortable stationary state, into the next stationary state where again we hunker down and minimize our entropy production. It is the positive feedback which drives our evolution. And so we mature, age, become aged, and head relentlessly towards maximum entropy, as we change in appearance and change in efficiency of internal body operations. Eventually we achieve maximum entropy, signaling a degree of disorder which makes it impossible to control life (the chemistry of life), and maintain the tension of life. And so we die. Death is the transition, from corporeal life to ethereal essence. Now the membrane, the boundary to the infinite, becomes permeable. It is a necessary condition, this condition of maximum entropy, for the tangible to cease functioning, for the essences to be released into Infinity and disappear. We have returned to the infinite reservoir of Infinity, where everything is, and everything was. At maximum entropy, everything returns, through dissolving boundaries, back into Infinity. We exist as singularities, stored information, discontinuities, temporary escapees from Infinity. To be returned to Infinity as potential information when conditions are right. Driven by entropy, Excess Entropy and Excess Entropy Production.

On earth, as time becomes infinite, things may settle down into a steady state, or into a fixed condition. But things may not settle down, may not converge at all, but may diverge and reach maximum entropy. On earth, liquid water freezing to solid ice, or gaseous steam condensing to liquid water, change entropy content significantly. Ice (low entropy), water (higher entropy), steam (highest entropy) trace the transition and evolution and phase changes from order, to disorder, to maximum disorder. We can calculate the entropy increase, from ice, to water, to steam. And where will the steam go as its entropy continues to increase, as we heat it more and more? Into Infinity?

Achieving maximum entropy is a randomizing process. It means landing in the most probable state, the universal attractor, with the most probable distribution of information energy, atoms, molecules, beings, things, worlds, universes.

Entropy characterizes symmetry. More symmetry, higher entropy. More randomization.

Entropy characterizes concepts, such as an expanding universe proceeds with increasing entropy.

The origins of life, the evolution of life, began as a self-organizing lurch to low entropy. Setting off a ticking clock of life and existence. When we achieve maximum entropy, it ends.

Life expectancies and the limitations of lifespan and the factors affecting it are entropy derived.

The orderly decay, the aging process, is entropy driven, towards maximum entropy.

Biological changes in aging, changes in brain weight, dietary manipulation of aging, are all entropy related.

Self-organizing communications networks are entropy driven.

Information flows in biological systems which self-organize themselves are entropy driven.

Dissipative structures are entropy driven.

Maintaining life far from equilibrium is entropy driven.

A gain in entropy means a loss of information.

Entropy tells us the direction things will go.

Entropy says we cannot take heat and convert it all to work.

Entropy says that when we mix things, there results more disorder, loss in information, more randomness.

Maximum entropy is when there is total randomness.

Entropy and Structure

Suppose we have two distinguishable gases, A and B, separated by a removable partition.

This is an ordered state since each gas is confined in only half of the box. If the partition is removed the gas molecules will mix until a homogeneous state is reached.

The entropy content of the mixed state has increased when the partition is removed. After the partition is removed, the molecules are more able to roam over the entire box. Without the partition there is greater uncertainty as to where a particular molecule will be. Thus with the state of higher entropy we associate the concepts of greater freedom, more uncertainty, and more configurationally variety. More uncertainty implies a higher probability of error. For example, if we had to guess where a particular molecule of gas A would be at a given time, the probability of error would be greater in the higher entropy state of mixed A and B molecules in the box without the partition.

Before the removal of the partition, we at least knew that molecule A was on a particular side of the box. Thus the ordering of a system (separating the gases A and B with a partition) implies lower entropy, which carries with it a certain reliability and smaller probability of error. We can summarize some characteristics associated with entropy as follows:

Higher Entropy	Lower Entropy
Random	Non-random
Disorganized	Organized
Disordered	Ordered
Configurational variety	Restricted arrangements
Freedom of choice	Constraint
Uncertainty	Reliability
Higher error probability	Fidelity
Potential information	Stored information

Infinity, Infinite Series

The series of numbers, $(1/2)^1, (1/2)^2, (1/2)^3, \ldots, (1/2)^n$ can go on forever, for $n = 1, 2, 3 \ldots$, towards very, very large values of n. Towards Infinity. What does it mean to say this infinite series goes on forever? One thing it means is that I can try to add all the terms, even counting all those numbers out there near Infinity. For example, let n equal one million, one trillion, and one zillion. The sum of this infinite series (adding up all the terms) can be shown mathematically to be equal to 1.

We say this series converges, and the sum of the Infinity of terms is equal to 1. Thus we have figured out how something behaves in the vicinity of Infinity. And we found structure and information. We are able to determine the sum; the sum converges in the vicinity of Infinity.

The series, $1, (1/2), (1/3), \ldots, (1/n)$ behaves in an opposite way, for the sum of an Infinity of these terms isn't a finite number at all, and we say the sum goes to Infinity. We say this series of numbers diverges. We have figured out how this particular something behaves in the vicinity of Infinity. We find it does not converge; it diverges, which means there is no structure and no information. This sum is indeterminate in the vicinity of Infinity.

We may conclude from these two examples that systems can behave differently in the vicinity of Infinity. Some things are clear there; some things are not. Some are determinant; others are indeterminate. We can see that $5/x$ goes to zero when x gets very large; very, very large: infinitely large. The result is determinant. It is also clear that $x/5$, the inverted from of $5/x$, behaves very differently. When $x = 10$, $x/5 = 2$, when $x = 100$, $x/5 = 20$, when $x = 1000$, $x/5 = 200$. But what is the value of $x/5$ when x gets very, very large, approaching an infinite number. When this happens $x/5$ also approaches an infinite number. We say it approaches Infinity (and is indeterminate). Dividing Infinity by 5 still gives Infinity. But what happens in this case, $x/2^x$, when x gets very large, approaching Infinity. Both numerator and denominator will get very large, each approaches Infinity. Yet there is something different at work in this example. We might sense that the numerator, x, will tend to be smaller than the denominator, 2^x. And even when we let x be a very large number, approaching an infinite value, this still will hold true. We can show mathematically that the 2^x Infinity will dominate the x Infinity, so that $x/2^x$ will go to zero (and is determinant) when x goes to Infinity. The lesson here is that apparently we can have two different magnitudes of Infinity.

What can we conclude from $2^x - x$, when x goes to Infinity? We have apparently Infinity minus Infinity. We are subtracting an infinite number from another infinite number. Is the result zero, Infinity, or something in between? Again, we can show that the two infinities are different, that the 2^x Infinity will dominate the x Infinity and the result is $2^x - x =$ Infinity when x goes to Infinity. We say $2^x - x$ is an indeterminate form until we figure out what the result is. Again the lesson to be learned is that not all Infinities are alike, not all Infinities are the same size, and that there are lesser Infinities which yield to the greater Infinity. That there is a super Infinity, within which all lesser Infinities are subsumed. Can we say the same for God? There may be many Gods, lesser Gods, who pale in comparison to the major Gods. Are the ultimate God and the ultimate Infinity the same?

Entropy, Structure, Information

Entropy yields evolution.

Entropy yields diversity.

Organizations, such as corporations, demonstrate entropy, structure, and information flow.

Countries and civilizations, likewise, show a hierarchical bureaucracy containing entropy, structure, and information flow.

The world, composed of discrete countries in ever changing alliances, dependencies, and strife, shows entropy, structure, and information flow.

Structure is a requirement for the storage of information, for without structure all the information available is simply potential information, theoretically available only when the system organizes itself.

We can calculate the entropy content of a corporate structure and conclude a certain behavior.

High entropy means more disorder than low entropy.

Living organisms and the so-called inanimate systems such as corporations, countries, and civilizations are similar dissipative organizations, living, growing, evolving, and dying as open systems, entropy producers, striving mightily to stay away from maximum entropy, going from one stationary state (at minimum entropy production) to the next stationary state (at another minimum entropy production), to the next, and the next, etc.

Living systems, you and I, as we proceed to maximum entropy (as we age) show signs that our chemistry of life, the life process itself, becomes slower and slower. We slow down as our internal structure disintegrates and our metabolism diminishes.

In the vicinity of death, in the vicinity of maximum entropy, in the vicinity of diminished structure, in the vicinity of minimum metabolism, there is barely enough energy and information to meet the simple stresses of ordinary life. A simple cold becomes a killing stress; a bump becomes a broken hip, which leads to diminished kidney function, affecting the heart, etc. There is not enough energy and information to return our bodies to a stable self. The smallest upset becomes a dangerous bifurcation point, where one of the paths leads to death.

Entropy and Infinity: Determinism, Randomness, and Uncertainty

I would like to be able to write an equation for my world (which Albert Einstein tried to do) and be able to say, if you do this, you'll get that. With this equation we could explain order in our lives, in the world, in our universe. We would like to be able to determine an output, resulting from an input, and be pretty sure what goes in is exactly what we intended. Our complicated living bodies, or the ambience of our world, or the behavior of our universe, may be deterministic and orderly (as Einstein suggested), or it may be chaotic and beyond our ability to understand it, and to predict events. Chaos may be a state approaching maximum entropy if there is no pattern discernable. And perhaps we've connected chaos, and our seemingly chaotic world, with a need to believe in God, for God is a sign, a symbol, of order and control. God gives us a deterministic chaos, in which the patterns and behaviors of our world are known only to God.

Albert Einstein said, "God does not throw dice," which means Einstein believed that God delivered order, but it was our responsibility to discover this order. To discover the patterns. Does something change by exponential growth or by arithmetic growth in going from here to there, from here to Infinity?

Something (x) happened to me now, at time, t, and we call that something x_t. Because of this, I became y_t. One moment later, at time $t + 1$, the something that happened is now x_{t+1}. Because of this, I became y_{t+1}. But in reality, we are all interdependent, and happenings x become y in complicated ways. I am who I am because of what was, what is, and what will be. You can write complicated equations which describe this, such as

$$x_{t+1} = x_t + y_{t+1}$$

which says that the happening later (x_{t+1}) depends not only on what happened before (x_t) but also on what I am to become later (y_{t+1}). Also,

$$y_{t+1} = y_t + k \sin x_t$$

which says what I will become later (y_{t+1}) depends not only on what I was before (y_t) but also on some complicated dependency on what happened to me before ($k \sin x_t$), where k is a constant and sin means the mathematical function, sine. We are all related, we depend on each other, and things change according to what you and I do separately and together. There is hidden order to our interactions. The equations describe it. But they also say things will appear chaotic until we understand the relationships. Which we may never be able to do. Religion, science, and art are in the same business: the business of discovering beauty and order and explanations for our behavior and existence and future in this world and beyond. We see patterns in history: circular (return from the future, back to the beginning and start again), helical (starting with the big bang and with an evolving and expanding universe, end with the big collapse), linear (we are traveling directly from here to there, wherever there happens to be), or non-linear (we are traveling from here to a bifurcation, where we might choose one path or another and proceed until the next bifurcation, etc.). We may be affected in our travels in so many simple ways as in this familiar example: the butterfly moving its wings in Maine sets up a very tiny current of air which affects the next air movement and this in turn affects something larger and more complex, until the weather in Ohio has been altered.

Infinity is being
Being is becoming
Becoming is evolving
Evolving is entropy
Entropy is aging
Aging is dissolving
Dissolving into Infinity.

Einstein in one of this thought experiments, postulated this scenario: a train is moving from left to right, passing one telegraph pole and heading towards another. When the train is halfway between the poles, lightning strikes each pole simultaneously. You are on the train holding a 90° mirror which allows you to see each pole with one look. You, the observer, are halfway between the poles, on the train, when the lightning strikes each pole simultaneously. Einstein proposed that you will see the lightning strike each pole at a different time. You will see the lightening strike to the pole you are moving towards, sooner. You will see lightning strike the pole you are moving from, later. Einstein theorized that, allowing that the speed of light is constant, the distance the light travels is less as we move toward the right hand pole. The light coming from the right pole (considered like a river, flowing and having some mass) contains the information that lightning has struck the pole and this light mass will arrive at the mirror and our eyes as we are moving to meet it. Einstein concluded his thought experiment by theorizing that time, the concept of time, is variable. Time, the duration of time, depends on our motion relative to the event from which the light is coming. And if we move at the speed of light, heading towards the event (the lightning strike on the right hand pole), then time actually stops, since it will take no time to see the event. The light hasn't had time to move away from the pole before we see it if we are traveling towards it at the speed of light. Do the speed of light and maximum entropy define our universe? Perhaps there are other systems beyond our knowledge with different standards for light velocity and maximum entropy, which define their distinctive definitions of time. A cosmological time, a variable time from one cosmos to the next. For us here on earth, we can distinguish between the past, present, and future by the passage of our time, which means, more fundamentally, by the passage of entropy. Living systems also mark the passage of time perhaps in terms of heartbeats. There is even a clock for living cell divisions, the maximum seems to be around fifty cell divisions over a lifetime.

The Maya of Central America believed that history would repeat itself every 260 years, a period called lamat. The cyclic pattern of time was a common feature in Greek thought. Every city once destroyed will be restored, people will reappear. Judeo-Christian tradition established linear (irreversible) time in western culture. Even Zeno's paradox addresses time and Infinity: Achilles chases a tortoise. During the time it takes Achilles to reach where the tortoise was, it has advanced a small distance. Achilles and the tortoise keep repeating the scenario. Will Achilles ever catch the tortoise? Is time, like space, capable of being infinitely divisible into ever smaller portions? Einstein believed the distinction between past, present, and future is only an illusion. He theorized that the faster the clock travels, the slower it ticks.

Irreversibility is connected with our sense of the passage of time: from the second law of thermodynamics we discover that entropy is intimately linked with time. Our earth's initial condition (the birth of the universe) started off small and highly compact, in a highly organized state with low entropy. Our passage of time corresponds then to increasing entropy. Self-organization also arises from second law considerations of our world, where things are evolving and changing, far from equilibrium. A chemical clock — chemicals which react and change color at regular intervals, is self-organization at work. Chemical products formed in these chemical reactions also participate and even catalyze their own manufacture. Also self-organization. Non-equilibrium thermodynamics bridges the divide between classical thermodynamics and Darwin's theory of evolution. Within a universe which is expanding, the arrow of time points from a highly organized big bang singularity to some maximally disorganized final state.

"Now one belongs entirely to nature, to time. Identity was a game. Memory has to be put aside, to travel to the curved ends of the universe, as light is said to do. I am in adolescence in reverse. Oh, I can comprehend a shutting down. A great power replacing me with someone else and with silence. My memories no longer apply to the body in which my words were formed. I don't know if the darkness is growing inward or if I am dissolving, softly exploding outwards, into constituent bits in other existences: micro existence. God is an immensity, while this death, which is me, this small, tightly defined pedestrian event. I am traveling now, and hearing myself laugh, at first with nerves and then with genuine amazement. It is all around me…" (Harold Brodkey, awaiting his imminent death from cancer, 1996)

Infinity. It is all about Infinity.

CHAPTER 8

On Time, Longevity and Entropy

The past no longer exists. The future has not arrived yet, but there is a need to plan for it. The present is reality.

Is time the prime reality as the Greeks believed, an independent and relentless power, destroying everything like an all-devouring monster? Time heals all wounds we say, implying an inherently good power; on the other hand, the degenerative wear-and-tear processes which yield ultimate death proceed with time as we are painfully aware. Time is associated with events like birth, death, growth, and decay.

Cause and Effect: Before and After

To attempt to understand time, we must explore the meaning of before and after and that syllogism which says because of A (which we observe) B was caused to happen, and therefore we conclude that B came after A. But how do you measure the event which is A? It is easy, we say. For example, it is a simple matter to observe the lightning flash in the sky and a moment later hear the thunderclap. So it is; but how long did it take for the light of the lightning bolt to reach our eyes? And from where did it begin? We know that sound waves travel more slowly than light — 1,000 feet per second compared to 186,000 miles per second — which means that the thunderclap will arrive at our ears later than the registering of the lightning flash in the eye, even though both are generated at the exact same instant. The closer the observer is to the lightning, the more simultaneous are the recordings of the occurrence. And as we all know, for one who is unfortunately in the immediate vicinity of the lightning, it is difficult for the unaided observer to tell which came first, the light or the sound.

The Form of Time

The early Greek philosophers raised the question of the structure of time: is time continuous or discontinuous; that is, can we divide time into an infinite number of ever-so-small intervals (continuous time) or is there some limited size of the discrete time interval, which implies a discontinuous march of time. Often time is considered a flow, as a river might flow, with a clearly defined direction. But rivers can be blocked, and major events can cause waterways to change direction and perhaps even reverse themselves. Can time be reversed? Is time therefore a physical reality — something tangible — or is it a psychological phenomenon, possessing no corporeal mass and related to and altered by our perceptions of it? Common sense might dictate that the rate of flow or movement of time is constant: a second is a second and an hour is an hour. But is a day a day and what does this mean? Physicists say that time marches on at a rate determined by circumstances, especially when huge cosmic distances or very high velocities are involved. If we define time as motion measured, such as "it took me 8 minutes to run the mile", then how are we to know when we are at the starting point and where is the end? Couldn't we take photographs at the start and end (when there is no motion), but how would we tell the time? On the other hand, if we photograph a person in youth and later in adulthood, it will be apparent that time has passed and, additionally, it should be easy to establish which is the earlier photograph.

Some say there are at least two theories of time: one that time is the product of the mind while physical things and places are timeless or, alternatively, time is not dependent on our thought processes; it exists as an independent reality, its presence real and unconnected with efforts to observe it. The first statement attempts to make the point that we have defined a concept. So we know an hour has passed when by definition the clock ticks off 3,600 intervals, and when the earth circles the sun we call that 1 day with 24 hours. But the second theory means the obverse that time exists in its own right as a tangible entity whose comprehension only awaits our ascending evolutionary development. At the proper stage in our history we shall, as Arthur Clark wrote in his story "2001", be able to fully understand what it is that we now call time. In the meantime we shall simply have to muddle along measuring and observing time's passage and building more and more sophisticated devices to indirectly record its existence.

The philosophers John Locke, David Hume and Rene Descartes considered the succession of ideas in the mind as the origin of our concept of time. We think, therefore we are — and in so doing define the ebb and flow of events, ordering them in relation to our mind's eye picture of our finite ambience and, by extension, the infinite universe. Or are we like someone on a boat in a river of time, floating through the countryside? Which is the true reality here? Is time the river or the permanent, stationary riverbank? And are there more rivers elsewhere, cutting through other countrysides? Perhaps our concept of past, present, and future has significance only in human thought and is in reality merely an anthropocentric idea, invented by humans to satisfy a need to be at the center of things.

The Measurement of Time

Time's measurement may be derived from a series of successive experiences, each anchored to a benchmark or point of reference. The procession from one benchmark to the next would define the passage of time, much as in a horse race when the lead horse passes first the quarter-mile pole, then the half-mile pole, and so on. But there is nothing new here, for the benchmarks still require us to know the time, suggesting that the concept of time is innate. As usual, and we should expect it by now, there is some evidence which seemingly contradicts the notion that time has an innateness: several kinds of animals — for example, birds, insects, and even humans — possess something called an internal or biological clock which is adaptable to changing environmental conditions. So in the spring when we move our clocks forward 1 hour in the parts of the United States where daylight saving time is practiced, the inhabitants adjust and go to bed an hour earlier than usual and awaken an hour later. Within a day or two we don't give this change a second thought. Seemingly, nothing has changed.

Returning for a moment to the imagery of time flowing as a river, suppose we in our mind remove the river shoreline and visualize time again as a great river but now self-contained, capable of flowing without confining shoreline boundaries. This immense stream without visible banks contains all there is, all that exists, floating on it. At certain moments, which here are specific locations in the river, each of us, a floating object, sinks, and disappears and hence no longer exists. Time flies, we say. What we really mean is time flows. All of us and everything that defines our existence are seen floating along, capable of holding our own for just so long before the river overwhelms our capabilities and we submerge.

It is the before-after logical chain which seems to be at the core of the discussion of time, for even the gravest cases of mental illness, patients utterly disoriented as to the time of day, can still tell whether a sound precedes or follows a light signal. The concept of time may have arisen when we became conscious of our reactions to certain sensations, so that first there were feelings (before) and then we acted (after). Or is it the other way? There is a belief that time is a series of events, that time as a concept was first founded on our ability to distinguish the sounds of these occurrences. But then the inevitable question is raised: what if we were insulated from sound?

Isolated events have no place in time, say some philosophers who propose that these events are not to be considered as pieces of wood floating in a time river. Instead, they say, why not compare time with the molecules of water which compose the river. Passing water molecules become the flow of time in this view. Time does not exist; it only flows. Thus the meaning of time has shifted from a flow of discrete events to a flow of a medium (a river) in which events takes place or have a place.

Though descriptions and definitions of time have not yet become focused, we still must work with time. To measure time, we need, first of all, to recognize or accept one simple paradigm; that is, time may be considered a sequence of events (though this sequence is not yet clearly ascertained). Next we choose a partial sequence which may be representative of the all-embracing one. We call such a partial sequence a clock. So the winding down of a spring in such a device approximates the 24-hour day, which in turn approximates the regularity of the movement of earth, sun, and stars. These heavenly body movements are assumed to be models of the behavior of the other galaxies and, hence, the universe as a whole. To say one's sense of time is relative, is to relate it to our own experience, where an interval of time in which many things happen is felt to be much longer than another interval of the same length in which almost nothing happens. True sometimes. But based on experimental evidence, we know that we must factor into these results the allowance that it is not only this information content but also the complexity of the information which affects our sense of time flow. Not only does the information content and its complexity guide or control our impressions, but, additionally, doesn't it make sense to expect that this availability of information requires on the observer's part an activated or receptive state. So for small children, up to about age 5, to tell them that something is to occur soon (in a week) needs some concretization. They understand better when we tell them that after they have gone to bed and awakened seven times, the week will be up. In earlier times we measured time by the rising and setting sun or the appearance of the moon. Time was experienced in the periodicity and rhythms in our lives as well as in the lifecycles in nature. Thus was invented the sundial, and Stonehenge and our present-day clocks with faces and hands. We designed devices to measure the duration of events, such as the so-called hourglass (in which sand flows through a constriction at a controlled rate) and the burning of a candle. Today we are a bit more sophisticated and utilize sidereal clocks which are based on the movement of stars and, by implication, relate time to our galaxy motion. There are light clocks which measure the distance traveled by light and pendulum clocks with their oscillating arms traversing distances at fixed-time intervals. Atomic clocks measure the vibrations in an atom of cesium.

The Direction of Time

There is only one time but many different types of clocks for measuring the partial sequence we designate as a time interval. Some liken the flow of time not to the flow of a river but to the flow of words. And if we cannot or do not think in words, does this imply no movement of time? Of course not. What is being suggested by the word-flow proponents is a return to the theory that time relates to our mental processes. Here we have the proposition that time was invented to account for a flow of a river (or torrent) of words and thoughts. If older people feel that the days of their youth were shorter than current days, is it true that time "flies" when we are busy; time "crawls" when we are bored? Then can the rate of flow of time be tied to the total number of events in the whole universe? Is the flow of time constant, infinite, indeterminant? One thing we believe we do know: the rate of flow of time cannot be negative. In other words, time's flows — forward and backward — are not equivalent. Time flows from past to future (or from earlier to later), and should we be able to reverse things and look at the future going to the past, we would not have an exact repetition of the process. For example, show a film in the correct sequence and then run the machine backward. It is true we will end up where we started but everyone and everything shown will be moving unnaturally and usually in violation of some accepted natural laws, such as dropped objects ascending unaided in violation of the law of gravity. A heavy body always falls down; heat always flows from the hot side of something to the cold. From these cases do we infer a direction for time: event A (the release of the heavy object) is always followed by event B (the fall of the object). Therefore A is the cause of B and A is earlier than B. What we are dealing with here is the observation that in nature there is a general tendency toward leveling; local differences tend to become spontaneously smaller and eventually disappear. The object, hot on one side and cold on the other, if allowed to stand untouched will show heat flow from the hot side to the cold side, the temperature difference will diminish with time, and eventually the entire object will be at the same temperature.

There is an inclination of every system to pass from a highly ordered, less probably state to a less ordered, more probably one. For example, such a highly ordered, less probable state might be a tank containing a gas under pressure. Open the valve on the tank and the gas molecules will spread into the surroundings and become mixed with the ambient air molecules. The new mixed condition is more usual or probable and less organized or ordered than is the initial high-pressure state. Given no interference, the tendency is always for the high-pressure gas to flow out of the tank. The more randomly things are distributed, the more likely is that state. The most probable state is characterized by maximum disorder or randomness. Spontaneous processes (the tank emptying, the heat flow from the hot to the cold surface, the dropped object falling to the ground) proceed in a way which may be considered to delineate the direction of time.

Having introduced the idea of events proceeding in one direction only, governed by seemingly immutable laws related to probability and order, this seems like the propitious moment (in time?) to bring up an arcane, artificial concept: entropy. Entropy can be defined as a measure of the randomness or disorder of a system and is related to its information content. Process proceeding spontaneously (remember the gas in the cylinder, the heat flow, and the fall of a heavy object examples) are said to go irreversibly for they cannot of their own volition return to their original conditions. Such one-sided events march from a more ordered to a less ordered state. They go in the direction from low probability to high probability. In entropy terms, we say they proceed in the direction of increasing entropy. So increasing entropy is associated with increasing disorder, which corresponds to the most probable state. Now we have another possible way of specifying the direction of the flow of time: unaided, natural processes head in the direction of increasing entropy, tending in the final analysis toward a state of maximum entropy. Even that ultimate of irreversible process — the course of human life, the aging process — proceeds with increasing entropy and its concomitant disorder.

Previously we said that time was tied to the observation of events. Time's arrow was seemingly connected to the before-after sequence, which allowed us to assess the direction of time flow. But there remains a nagging question which must be faced, having to do with whether it might be possible for observers to see a series of events in different order. Is the direction of time the same throughout the universe? What we mean by a reversal of the direction of time is that the later events take place sooner, so that for the vase which fell and broke — now the pieces are observed to fly upward and join to form the original vase. This is clearly impossible to accomplish unaided, at least as far as our present knowledge dictates. A partial reversal of the direction of time would seem unlikely (only the vase and nothing else in our world may repair itself spontaneously). It is all or nothing it would seem; either the entire universe is capable of time reversal or nothing is. If time reversal could occur, all laws of nature would keep their validity and the whole world would run through the exact same states as before, only in the reverse order of succession. Somehow this seems logically impossible. But perhaps it is wrong to dismiss this question of reversibility so facilely. If we think of time reversal as a reversal of the motion of elementary particles such as electrons and other parts of atoms, it may indeed be possible for these particles to pass through exactly the same states as before. In other words, we say there is the possibility of microscopic reversibility, but macroscopic systems (the whole which is composed of these elementary particles) such as the falling object and the gas in the cylinder cannot be made to run unaided through the exact same sequence of states as before, but in the reverse order of succession.

Time's Duration

Time is relative in some respects. Here relative means being dependent on something else. The theory of relativity from physics holds that the duration of an interval between two events is dependent upon the state of motion of the observer. To an observer at rest this interval would be longer than it would be to an observer in motion. Still another difficulty in ascertaining the direction and duration of time has to do with the problem of establishing simultaneity. In other words, how can you tell when two events occur exactly at the same moment. For example, the simultaneity of events on two distant stars is difficult to establish. On earth it would seem easier to do since the distances are smaller, and the light or electronic signal which conveys our information travels at such astronomical velocities. But a happening on a star might be observed many years after its occurrence. Not only that, but the velocity of light is not always known exactly, for the light rays can be deflected by a heavy body such as a star.

The theory of relativity says that temporal or time relationships and the events we measure do not exist independently by themselves, but are linked temporarily to an observer. The "instant" when an event occurs is localized and can be different in different places. From physics we know that a clock in motion runs slower than a clock at rest. Now we have added the complication of the motion of the clock and the observer to all the other problems in establishing when events occurred in distant stars — and in defining time and its direction (if it does indeed have a direction). Perhaps what we measure with a clock isn't all that accurate anyway, but is specious. Consider the case of the hypothetical space traveler. The traveler departs from earth in a rocket at or near the speed of light and on the return appears to have aged much less than those who remained on earth. The explanation of the puzzle is usually given in terms of the difference in the time interval experienced by the space traveler and the earth dweller. We are measuring things with two non-synchronous clocks; one goes with the rocket at or near the speed of light and the other remains on earth and moves at a lesser velocity. The two clocks tick off different time intervals: the cumulative time interval for the space travelers is less than for the earthbound.

As speed increases, time seems to slow. If we are able to achieve speeds equal to the speed of light, equations predict that time would finally stand still. Recently new elementary particles have been discovered and labeled tachyons. These tachyons are believed to move faster than the speed of light. If this is true, can we say that for the tachyons time flows backward since they travel faster than the limiting velocity — the speed of light — where time stands still. This is probably an absurd conclusion to draw so there must be other possibilities. One alternative hypothesizes that tachyons are entirely different particles than the usual ones and for them the limit is still the speed of light but they approach this boundary from the other side, coming down in velocity toward the speed of light. Most particles and systems we have experienced approach the speed of light from below.

There is still the question of whether time is continuous or discontinuous. If time is continuous, it can be divided into an infinite number of infinitesimal intervals. Or is time discontinuous and hence can only consist of discrete atoms or quanta of finite duration. If time is continuous, is it like a line; the line, we say, is composed of an infinite number of points. If time were discontinuous and composed of discrete moments, do the moments touch? How long is each moment? What do we call the space between moments if they don't touch? Time periods smaller than 10^{-22} seconds cannot be measured today and as a consequence we have no idea whether a time moment can be less than this. The smallest increment of space which can be measured, a so-called space quantum, consists of 10^{-15} meters where a meter is 100 centimeters or a little more than a yard in length. If both time and space consist of quanta, perhaps fuzzy in appearance so that it is difficult to tell where one ends and the next begins, only extremely close observations could yield a flow of time which appeared to be discontinuous. Most events in everyday life are measured no closer than a hundredth of a second and time therefore would seem continuous because of our inability to distinguish the exact boundary between earlier and later quanta of time. By assuming the quanta to be fuzzy and fading into each other, we can neatly avoid the need to choose which model of time we prefer: the continuous or discontinuous one. We can visualize particles gliding continuously from one time quanta into the next. These quanta can be compared with droplets of liquids. Droplets can coalesce to form a continuum, which is one of the proposed models for the flow of time.

Cyclic Time

Of course, time could also be cyclic and if you believe this, the universe will have to retrace its past steps and go through the same series of events. The proponents of the cyclic theory ask, why must we assume that time had a beginning as do the linear time thinkers. Why would the first event in our history occur? Why should time have an end? Shouldn't time be infinite at both ends of its boundary, in the past and future? These cyclic time believers say there is nothing in our experience which supports the view that there has ever been a first event or that there will be a last. Hence why couldn't time be cyclic, returning to its original state again and again. During one cycle we trace out a history and return to the beginning. Then the history unfolds again. In essence there is no beginning or end in this model. Thus, if the universe were awakened from its deathlike rest once by a big bang, why not a second time and third time and so on. If moments of time have no identity in themselves but owe their character to the events which take place at them, the cyclic theory of time implies a cyclic theory of events. But with cycles, if a particular moment begins a cycle, it can be perceived as being before every other moment. It becomes a matter of definition: where do we choose to begin the cycle? If we wish, we could arrange the cycle so that a particular moment comes after every other moment. But with this latitude comes the possibility that an effect could precede its cause, allowing such unreasonable conclusions as death could precede birth. With this logical difficulty to be faced, is it any wonder that the cyclic theory of time has few followers?

In summary, time has to do with events. Time is a primitive relationship; it is a product of the mind but based on the reality of physical bodies and what happens to them. Time is infinite and may or may not be continuous. Temporal or time relationships are between changes of the state of bodies. Time has only one dimension. One can travel to another point in time in one way only, by a completely determined series of intermediate points in time: from past to present to future.

Longevity and Entropy

Some life-science researchers define an organizational entropy and show that

$$S_{org} = R(\ln u/\rho) + \text{constant}$$

where

S_{org} = organizational entropy

R = gas constant

ρ = maturation rate for a definitive stage of development
     of the organism

u = mean metabolic rate for the same stage

u/ρ = measure of the energy cost of carrying the development
     of an organism from one defined stage to another

A lower rate of entropy production permits the organism to live longer and do more metabolic work. The rate of aging and maturation are apparently related to the rate of entropy production of the whole system: aging is related not only to how much metabolic work is performed but to how well the work is done, in entropic terms. Maximum entropy may correspond to death. If the death of an organism is viewed as the state characterized by maximum entropy, we ought to determine the entropy production during the lifetime of an organism. Comparison of the total lifetime entropy production for different organisms should be enlightening, and from these figures perhaps some inferences can be made concerning life expectancy.

Hershey defined an "organic entropy" for homeothermic systems as $\Delta S = \Delta Q/T$, where ΔQ is the basal heat per unit weight per unit time given off by the organism, T is the isothermal temperature, ΔS is the organic entropy. He calculated the integral, organic entropy over the lifespan of various animals and found the calculated lifetime summation was of the same order of magnitude for the animals examined. Calloway suggested that is not the sum total of energy or entropy, but the approach to a lower critical level, which signals the onset of death. He puts this figure as 0.833 Kcal/Kg hr for various animal species and suggested perhaps that this represented a universal characteristic for living tissue.

The human body is a highly improbable and complex system of organs and tissue which involve about 60 trillion cells. While the body has a natural mechanism for restoring cells and combinations of cells to their proper states and functions, the process never results in perfect restoration and alignment of all the cells that make up the body. Thus some imperfection always remains. At first these imperfections are unnoticeable on a macroscopic level; with time, however, more and more cells are not restored to their original configuration and position. These imperfections thus gradually accumulate until the critical level of imperfections occurs and the entire system collapses.

The aging process is always accompanied by the continuous development of imperfections. The lifespan of human beings depends on their environment, heredity, life style, nutrition, and mental state. Under certain conditions, a person may live over 100 years. From one point of view, the essential feature of living organisms is their ability to capture, transform, and store various forms of energy according to specific instructions carried by their individual genetic material. Living organisms need to acquire energy to do biological work for maintaining their life. Some biological work is easily understood. For example, the heart has to work as it pumps blood. Some biological work is less obvious: for example, the work done by the intestine in absorbing foodstuffs. Work is also done when there is a high level of genetic and nervous activity. To function effectively, biological systems must be programmed to acquire information about the internal and external environment. This type of work controls energetic processes, organizes bio-structures, and controls the energy needed for fast response to stimuli.

Energy is useful when it can be converted into work. Entropy laws state there is always a certain amount of energy which changes into a lower-quality form and becomes less available to do work. This is not caused by an inherent problem of design in any particular engine or process; it is a law of the universe which applies, as far as we know, from the smallest atom to the largest galaxy.

Let us denote S as the entropy of the human system. We can write the following equation:

$$S = \log f(N, E)$$

where

$N =$ Number of particles in the human system
$E =$ Internal energy of the human system

The form of this equation tells us that entropy increases as the number of cells and the total energy within the body increases. Thus as the body grows beyond some sort of optimal configuration, more disorder occurs within it. Also as we eat, we increase our total energy content and more disorder again occurs.

From the many definitions of entropy related to disorder, probability of certain states of systems, presence of uncertainty, and topology, we can draw some inferences about the relationships between the entropy of a physical system and other physical quantities.

1. An increase in internal energy of a system increases its entropy.
2. A system will decay faster if insufficient work from its surroundings is applied to the system or insufficient internal work is done inside the system.
3. The passage of time will cause the entropy of a system to increase automatically unless adequate low entropy is supplied.
4. As the entropy of a system increases, the degree of random activities within the system will increase.
5. As a system's entropy increases, its energy become less available for doing useful work.

Isolated systems as well as equilibrium systems attain a maximum disorder. All equilibrium structures are stable to small perturbations. Since the turn of the century, phenomena have been identified which seemed to operate differently. Among them were a periodic precipitation phenomenon where a concentrated salt solution such as lead nitrate diffuses into a lyophilic gel such as agar-agar containing potassium iodide. A precipitation of lead iodide forms discontinuous bands (Liesegang rings) parallel to the diffusion front. Other examples of systems initially far from equilibrium are the Belousoff-Zhabotinsky reaction (bromination of malonic acid) and Bernard cells (thermal stability in horizontal layers of a fluid which is heated from below, a cellular convection structure).

The departure from classical equilibrium thermodynamics started with Onsager in 1931. He analyzed non-equilibrium situations not too far from equilibrium by showing that linear relations hold for the association of fluxes (heat and matter) with thermodynamics forces (temperature and chemical potential). Onsager coined the expression "the principle of least dissipation of energy" which applied to stationary states in the linear region. This statement implies that a physical open system evolves until it attains a stationary state where the rate of dissipation is minimal. Prigogine proved this implication in 1945 and called it the principle of minimum entropy production.

Prigogine and coworkers asserted that near-equilibrium stationary states with minimum entropy production are uninteresting (the entropy production term was a Lyapounov function implying that the stationary states are always stable). Thus any spontaneous fluctuation arising in this system regresses in time and disappears. Such a system near equilibrium cannot evolve spontaneously to new and interesting structures. Prigogine wrote that systems far from equilibrium with nonlinearities (autocatalytic or feedback loops) can evolve spontaneously to new structures. He referred to equilibrium or near-equilibrium states as the thermodynamic branch, whereas the new structures are called dissipative structures. Beyond the thermodynamic branch, physical systems show a new type of organization. The dissipative structures can be maintained only through a sufficient flow of energy and matter. The work required to maintain the system far from equilibrium is the source of the formation of order. Fluctuations play a crucial role near the point of instability; they become large and are built up by the nonlinear behavior of the system into dissipative structures.

Glansdorf and Prigogine in 1971 published a monograph on the theory of stability in the thermodynamic branch where,

$$\Delta S = \delta S + 1/2\delta^2 S + \cdots ,$$

and $1/2\delta^2 S$ is the Excess Entropy.

If $\left(\frac{\partial}{\partial t}\right) \delta^2 S > 0$, then the Excess Entropy is a Lyapounov function and the state is stable. A sufficient condition of instability is $\left(\frac{\partial}{\partial t}\right) \delta^2 S < 0$. This analysis is limited to small fluctuations since higher-order terms in the ΔS expansion have been neglected.

The possibility of building order through fluctuations under extreme non-equilibrium conditions has implications for bio-morphology, embryology, and evolution. The ideas of self-organization are not limited to chemical and biological systems. Other applications are in fields such as population dynamics, meteorology, economics, and even the urban existence of a big city which can survive only as long as food, fuel, and so on flow in while wastes flow out.

In thermodynamics, the second law appears as the evolution law of continuous disorganization or the disappearance of structure introduced by the initial conditions. In biology or sociology, the idea of evolution is, on the contrary, related to the increase in organization, resulting in structure whose complexity is ever increased. Thus the classical thermodynamic point of view indicated that chaos is progressively taking over, whereas biology points in the opposite direction. Are there two sets of physical laws that need to be involved to account for such differences in behavior? Prigogine said there is only one type of physical law but different thermodynamic situations: near and far from equilibrium. Destruction of structure is the typical behavior in the neighborhood of thermodynamic equilibrium. Creation of structure may occur when nonlinear kinetic mechanisms operate beyond the stability limit of the thermodynamic branch.

Prigogine's ideas may be summarized as follows.

1. Closed systems and linear systems, in general, close to equilibrium, evolve always to a disordered regime corresponding to a steady state which is stable with respect to all disturbances. Stability can be expressed in terms of a minimum entropy production principle.
2. Creation of structure may occur spontaneously in nonlinear open systems maintained beyond a critical distance from equilibrium. The system evolves to a new regime, an organized state (dissipative structure). These new structures are created and maintained by the dissipative entropy-producing processes inside the system.
3. Thermodynamic equilibrium may be characterized by the minimum of the Helmholtz free energy, $F = E - TS$, where E is the internal energy, T is the absolute temperature, and S is entropy. Positive time, the direction of time's arrow, is associated with the increase in entropy.

Isolated or closed systems evolve to an equilibrium state characterized by the existence of a thermodynamic potential such as the Helmholtz or Gibbs free energy. These thermodynamic potentials and also entropy are, according to Prigogine, Lyapounov functions, which means they drive the system toward equilibrium in the face of small disturbances.

4. Entropy production is $dS/dt = \sum_\rho J_\rho X_\rho \geq 0$, in the neighborhood of equilibrium. This is also the basic formula of the macroscopic thermodynamics of irreversible processes. Here J_ρ is defined as the flux of the irreversible process (chemical reaction, heat flow, diffusion) and X_ρ is the generalized force (affinity, gradient of temperature, gradient of chemical potential). Near equilibrium there results linear homogeneous relations between flows and forces (Fourier's law; Fick's law). A second principle valid near equilibrium is the theory of minimum entropy production for stationary steady states. This occurs in the linear range when boundary conditions prevent entropy production from becoming zero (thermodynamic equilibrium); rather the system settles down to a nonzero minimum level.

5. Closed equilibrium states are at maximum entropy, and if perturbed, the entropy can be expressed as $S = S_0 + \delta S + \frac{1}{2}\delta^2 S + \cdots$, where S_0 is the equilibrium entropy value. However, because the equilibrium state was at a maximum entropy value, δS vanish and stability is given by the sign of the second-order $\delta^2 S$ term (a Lyapounov function which has a damping effect on the fluctuation).

6. Prigogine gives a specific example to illustrate the behavior of chemical systems far from equilibrium. If the kinetics are as follows:

$$A \to X$$
$$2X + Y \to 3X$$
$$B + X \to Y + D$$
$$X \to E$$

then the rate of appearance of components X and Y is of the form

$$dX/dt = A + X^2Y - BX - X$$
$$dY/dt = BX - X^2Y$$

With diffusion, these equations become

$$\partial X/\partial t = A + X^2Y - BX - X + D_x \partial^2 X/\partial r^2$$
$$\partial Y/\partial t = BX - X^2Y + D_Y \partial^2 Y/\partial r^2$$

where D_x, D_Y are the diffusion coefficients.

7. Closed systems evolve toward an equilibrium state characterized by a zero entropy production rate and maximum entropy content. Open systems not far from equilibrium are drawn toward a stationary or steady state where entropy production has achieved a minimum. These open systems in the stationary state continue to generate entropy linearly with time. A living open system such as the cell or the human body may not achieve a stationary state. It is not clear whether the living system is near or far from equilibrium. Instead of tending toward a stationary state, living systems may evolve toward death which may be characterized by the organism slipping below some critical level of entropy production. Below this critical level, the system simply cannot support life.

Entropic Analysis of the Living Human System

The aging process has been studied for a long time, but no theory has been presented that provides all the answers to the "hows and whys" of aging. Since entropy seems to be one of the important variables in nature which may at times parallel the direction and irreversibility of time, the entropy concept appeals to many as a powerful tool in the understanding of the aging processes.

Entropy is a vital concept for our time. It has direct relevance in the study of shrinking resources, increased pollution, and a greater sense of social responsibility, all of which characterizes our present and shapes our future. It has been applied to various closed physiochemical systems for about 120 years. One form of the second law of classical thermodynamics is $dS/dt \geq 0$ (isolated systems) which means the entropy of an isolated system never decreases.

The applicability of the entropy law to biological processes is a thorny problem in physical/biological reductionism. When the question was submitted to an international conference at the College de France in 1938, it caused much acrimonious debate, but no agreement could be reached. The nub of the argument was that the second law of thermodynamics applies only to isolated systems: systems which exchange neither matter not energy with their environment. Living systems, in contrast, must exchange matter and energy with the environment in order to survive.

Prigogine formulated an extended form of the second law of thermodynamics which applies not only to isolated systems but also to open systems, that is, systems which exchange both matter and energy with the environment. Subsequently, Prigogine and his colleagues divided the total entropy variation into two parts: the entropy flow due to irreversible processes within the system and the entropy exchanges with the surroundings. The internal entropy production is always zero or positive, but the external entropy flow can have any sign.

The living system is essentially an open system because it maintains itself by the exchange of matter and energy with the environment and by the continuous building up and breaking down of its internal components.

Let $d_e S/dt$ designate the rate of entropy exchanged with the surroundings during time, dt, and $d_i S/dt$ represent the rate of production of entropy within the system during the same time. For the total rate of variation in entropy of the system, dS/dt, we can write

$$dS/dt = d_e S/dt + d_i S/dt$$

The internal rate of production of entropy is related to the irreversible phenomena that occur within the living organism (chemical reactions, heat transfer, mass transfer, and so forth), and is nonnegative, that is

$$d_i S/dt = 0 \quad \text{(reversible processes)}$$

$$d_i S/dt > 0 \quad \text{(irreversible processes)}$$

In other words, irreversible processes create entropy. The external entropy flow rate, $d_e S/dt$, is related to the transfer of energy and matter in and out of the system. The sign of $d_e S/dt$ varies and depends on the direction of exchange. As a result, dS/dt will be larger or smaller than zero depending on the importance of $d_e S/dt$. (It can be shown that in a living organism, the rate of internal production of entropy due to its metabolism surpasses by far the $d_e S/dt$ terms).

For isolated systems (internal energy and volume are constant and no mass or heat is exchanged) we obtain

$$d_e S/dt = 0 \quad \text{(isolated systems)}$$

$$dS/dt = d_i S/dt \geq 0 \quad \text{(isolated systems)}$$

For a system with an internal chemical reaction of velocity, V, and affinity, A, there is a production of entropy per unit time which is expressed by

$$d_iS/dt = AV/T$$

where T is the absolute temperature; and affinity, A, will be defined shortly. If there are several chemical reactions, the resulting production of entropy per unit of time is a summed contribution, that is

$$d_iS/dt = \sum_j A_jV_j/T$$

where the affinity, A_j, is given by

$$A_j = -\sum_j \upsilon_{kj}\mu_{kj}$$

and υ_{kj} = stoichiometric coefficient of the jth chemical species in the kth chemical reaction; μ_{kj} = chemical potential of the jth chemical species in the kth chemical reaction.

There has been interest in this concept of entropy production as applied to stationary states not in equilibrium. This is possible when a balance is struck between internal production and external transfer of entropy with neither term being necessarily zero. Processes likely to demonstrate this are thermal diffusion and the Knudsen effect, among others. For biological systems, one can often assume that a living organism is in a state approaching the stationary state that is

$$(dS/dt)_{t_f} = 0$$

where t_f is the time when the stationary state is achieved. For biological systems, the stationary state may be equivalent to senile death.

Previous analyses of living systems from a thermodynamic point of view have led to these hypotheses.

1. In the course of living — and of evolution — there is and has been a tendency to modify the internal entropy production rate, d_iS/dt (the sum of metabolic processes), in such a way that

$$d/dt(d_iS/dt) \leq 0$$

In other words, the rate of internal entropy production is continuously decreasing and may be tending toward a minimum.

2. The living organism may in the course of its life evolve toward a stationary state (senile death) which may correspond not only to $(dS/dt)_{t_f} = 0$ but also to a minimum of the total rate of change of entropy, that is

$$d/dt(dS/dt)_{t_f} = 0$$

The human body is an open system, continuously exchanging mass and energy with its surroundings. There are many internal biochemical reactions, and therefore heat and mass transfer are occurring internally. All these processes are irreversible and interact with each other. We assume that the major source of internal entropy production is related to the chemical reactions inside the organism.

Let S designate the total entropy content of the human system, with $d_e S/dt$ representing the differential entropy rate of change with the surroundings due to heat and mass transport, and $d_i S/dt$ being the differential rate of internal entropy production (for internal biochemical reactions).

Then

$$d_i S/dt = d_e S/dt + d_i S/dt$$
$$d_i S/dt > 0 \quad \text{(irreversible processes)}$$
$$d_i S/dt = \sum_j A_j V_j/T$$
$$(dS/dt)_{t_f} = 0 \text{ in the vicinity of death}$$
$$d/dt(dS/dt)_{t_f} = 0 \text{ in the vicinity of death}$$

The chemical affinity, A_j, can be related to the Gibbs free energy, G, by

$$A = -(\partial G/\partial \xi_j)$$

where, ξ_j = extent of the jth chemical reaction, T = absolute temperature.
Using

$$G = H - TS$$

and

$$(\partial H/\partial \xi_j) = -(r_j)_{T,P}$$

where, $(r_j)_{T,P}$ = heat of reaction of the jth reaction at constant temperature and pressure H = enthalpy and by combining equations we obtain

$$A_j = (\partial H/\partial \xi_j) + T(\partial S/\partial \xi_j)_{T,P}$$
$$= (r_j)_{T,P} + T(\partial S/\partial \xi_j)_{T,P}$$

In textbooks on classical thermodynamics, it is shown that the term, $T(\partial S/\partial \xi_j)_{T,P}$, can often be neglected in comparison to $(r_j)_{T,P}$. We assume this for the living system. Thus from the equations, we can get

$$\frac{d_iS}{dt} \cong \frac{1}{T} \sum_j (r_j)_{T,P} V_j = \frac{-1}{T} \sum_j (\partial Q/\partial t)_{T,P}$$

where $(\partial Q/\partial t)_{T,P}$ is the rate of internal heat generation due to chemical reactions and V is the reaction velocity. By thermodynamics convention $\partial Q/\partial t$ will be negative. Thus the internal entropy production of a living homoeothermic organism might be measured by its basal metabolism, as recorded by calorimetry.

Energy Metabolism Measurement

An important advance in physiology was the demonstration that the amount of energy liberated by the catabolism of food in the body is the same as the amount liberated when food is burned outside the body. The energy liberated by catabolic processes in the body appears as external work, heat, and energy storage.

The amount of energy liberated per unit of time is called the metabolic rate. Energy is stored by forming energy-rich compounds. In an individual who is not moving (no external work) and has not eaten recently (no energy storage), essentially all the energy generated appears as heat. Thus in a resting, fasting state, the metabolic activity can be measured as the rate of heat transfer from the body to the environment. To make a comparison of the metabolic rate of different individuals and different species, metabolic rates are usually determined at as complete mental and physical rest as possible, in a room with a comfortable temperature, 12 to 14 hours after the last meal. The metabolic rate determined under these conditions is called the basal metabolic rate (BMR). Actually the rate is not truly basal; the metabolic rate during sleep is lower than the basal rate. What the term basal denotes is a set of widely known and accepted standard conditions, listed.

1. The subject has not been exercising for a period of 30 to 60 minutes prior to the measurement.
2. The subject is in a state of absolute mental and physical rest, but awake (the sympathetic nervous system is not overactive).
3. The subject must not have eaten anything during the last 12-to-14 hour period prior to the measurement. (Protein needs up to 12 to 14 hours to be completely metabolized.)
4. The ambient air temperature must be comfortable, 62 to 87°F (which prevent stimulation of the sympathetic nervous system).
5. The subject must have a normal body temperature of approximately 98.6°F.
6. The pulse rate and respiration must be below 80 beats per minute and 25 cycles per minute, respectively.
7. The subject should wear a loose-fitting gown to keep the same experimental conditions each time.

As discussed previously, for the human in a basal state, essentially all the energy output from the catabolism of food in the body appears as heat and the rate of internal production of entropy related to its metabolism surpasses by far that connected with others causes of irreversibility. So we may calculate the internal entropy production rate by

$$d_i S/dt = BMR/T_B$$

where BMR represents the basal metabolic rate of the human subjects (corrected for water evaporation from the skin and internal surfaces) and T_B denotes the constant body temperature.

In principle, energy production in the living organism could be calculated by measuring the products of the energy-producing biological oxidations — carbon dioxide, water, and the end products of protein catabolism — or by measuring the oxygen consumed. These are indirect calorimetric methods. It is difficult to measure the end products, but measurements of oxygen consumption are relatively easy. (Since oxygen consumption keeps pace with immediate needs, the amount of oxygen consumed per unit of time is proportional to the energy liberated.)

One problem with oxygen consumption as a measure of energy output is the fact that the amount of energy released per mol of oxygen consumed varies according to the type of material being oxidized. The average value for energy liberation per liter of oxygen consumed is 4.82 Kcal, and for many purposes this value is accurate enough. More accurate measurements require data on the foods being oxidized. Such data can be obtained from an analysis of the respiratory quotient and the nitrogen excretion.

Respiratory quotient (RQ) is the ratio of the volume of carbon dioxide produced to the volume of oxygen consumed per unit of time. It can be calculated for reactions outside the body, for individual organs and tissues, and for the whole body. The RQ of carbohydrate is 1.00 and that of fat is about 0.70. This is because oxygen and hydrogen are present in carbohydrate in the same proportions as in water, whereas in various fats extra oxygen is necessary for the formation of water. For example:

$$C_6H_{12}O_6 + 6O_2 \rightarrow 6CO_2 + 6H_2O$$
Glucose

$$C_{12}H_{22}O_{11} + 12O_2 \rightarrow 12CO_2 + 11H_2O$$
Lactose

$$C_6H_{10}O_5 + 6O_2 \rightarrow 6CO_2 + 5H_2O$$
Glycogen

are typical oxidation reactions of carbohydrates. But

$$(C_{15}H_{31}COO)C_3H_5 + 27O_2 \rightarrow 19CO_2 + 18H_2O$$
Glycerol tripalmitate

$$(C_{17}H_{33}COO)C_3H_5 + 29\tfrac{1}{2}O_2 \rightarrow 21CO_2 + 19H_2O$$
Glycerol trioleate

are typical oxidation reactions of lipids. Determining the RQ of protein in the body is a complex process, but an average value of 0.82 has been calculated. RQs of some other important substances are glycerol (0.86), acetoacetic acid (1.00), pyruvic acid (1.20), and ethyl alcohol (0.67).

The approximate proportions of carbohydrate, protein, and fat being oxidized in the body at any given time can be calculated from the carbon dioxide expired, the oxygen inspired, and the urinary nitrogen excretion. However, the values calculated in this fashion are only approximations since the volume of carbon dioxide expired and the volume of oxygen inspired may vary with metabolism and respiration. Therefore measuring basal metabolic heat given off by an oxygen consumption method can be inaccurate since it is based on a number of tenuous assumptions and uncertain data.

The energy released by test-tube combustion of foodstuffs can be measured directly by oxidizing the material in a bomb calorimeter (a metal vessel surrounded by water, all inside an insulated container). The food is ignited by an electric spark. The change in the temperature of the surrounding water is a measure of the calories produced. Measurements of the energy released by living organism combustion of compounds are much more complex, but large calorimeters have been constructed which can physically accommodate human beings.

The caloric values of common foodstuffs, as measured in a bomb calorimeter, are 4.1 Kcal/gm of carbohydrate, 9.3 Kcal/gm of fat, and 5.3 Kcal/gm of protein. In the body, similar values are obtained for carbohydrates and fats but the oxidation of protein is incomplete, the end products of protein catabolism being urea and related nitrogenous compounds in addition to carbon dioxide and water. Therefore the caloric value of protein in the body is an estimated 4.1 Kcal/gm.

Since in the basal state the metabolic rate can be measured by the rate of heat transfer from the body to the environment, Hershey designed and constructed a whole-body calorimeter to measure the basal metabolic rate of elderly human subjects. The idea is to determine the rate of heat transfer from the body to the whole-body calorimeter. From the physical properties of the inlet and outlet airstreams, he calculated this heat transfer rate based on previously established calibration constants and a heat balance on the whole-body calorimeter.

We see generally decreasing values for the Basal Metabolic Rate (BMR) as we age. Some believe that there is a minimum level of the BMR, below which (on average) we cannot support life. For males, this death age is about 85 years and for females it is about 100 years. In other words, in general, the male population would have a lifespan potential of 85; for females it would be 100 years. This is the expectation, assuming we do not die prematurely, in a car accident, of a serious illness, or other stresses which diminishes our potential.

Some believe that our lifespan potential is dictated by our lifetime energy availability. If we had a more or less fixed amount of energy available to be consumed (or generated) over our lifetime, then we could, in theory, calculate the energy we consume or generate during our lifetime, and when it approaches the potential for our gender (male or female), and species (human), we would approach senile death.

For a metabolic rate versus age curve the area under the standard curve would be the potential lifetime energy expenditure. For an individual, the area under the curve could be extrapolated until the area under the individual's curve equals the area under the standard (average) curve. Thus the individual's curve would end at the death age, the longevity projection when the two areas are equal.

The Basal Metabolic Rate and Aging

The major influences on lifespan, assuming we are reasonably healthy, are our genetic makeup and internal chemistry. Beyond this, the external environment, such as pollution, radiation, the nature of the food ingested, our physical condition, etc., are also factors in determining longevity potential. These chemical reactions give off heat (much as an atomic bomb does, but much less magnitude, of course). It is this heat transfer which gives us a clue as to the vitality of our living system. It is this heat evolution, under prescribed resting conditions, that we call the Basal Metabolic Rate (BMR), a quantity known to decrease with age. We can convert this BMR into entropy units. Most people show what we call type I BMR behavior. People who display the type I history are living "normal" lives, which means being free from significant medical stresses such as prolonged illness. The BMR of type I people decreases steadily with age until death. Another BMR pattern observed is called type II, where a subject begins in a type I mode, but then shows a BMR which increases. The subject may die while tracing out a type II line, whose duration may be 1–5 years. (The stress causing this aberrant behavior is defined as a "killing stress".) Obviously the virulence of this killing stress depends on the age of onset and its intensity. The subject may recover from type II behavior, through the natural healing powers of the body or by medical intervention, and resume the more normal pattern called type III behavior. This person has recovered from the type II stress, so obviously it was not a killing stress. The lifespans for type III subjects tend to be less than type I but there are a variety of possibilities. For example, the stress might have been promptly diagnosed and successfully treated suggesting perhaps an extended longevity potential. The efficacy of the treatment is reflected in the resumption of the normal downward slant of the BMR. Factored into this history in addition to the promptness of the diagnosis is the severity of the stress and the nature of the treatment. (Some obese patients might initiate weight loss and exercise programs.) Obviously diagnosis does not guarantee recovery; for example, some of the type II cases might arise from untreatable cancer.

Factors which may increase the BMR:

Condition	Comments
Hyperthyroidism	Can double BMR in severe cases
Prolonged overfeeding	11 out of 16 studies report increases
Physical trauma (burns, injuries, etc.)	Degree of change related to the severity of the trauma
Deep body temperature rise	BMR increases 12% per degree change outside 27–34°C zone
Anxiety	Increase can be significant
Diabetes	Severe diabetes can increase BMR
Growth hormones	Increases observed in subjects given injections
Hypertension	Increases observed
Cardiovascular disorders	Increases observed in aortic stenosis, congestive heart failure
Endocrine disorders	Increases observed in acromegaly, Cushing's disease
Leukemia	Can increase BMR by 120%; increase related to degree of disease advancement
Multiple myeloma	Increases of 14–60% observed
Carcinoma	Increases of 6–57% observed
Various cancers	Increases above normal observed

Excess Entropy (EE) and Excess Entropy Production (EEP) Driving Forces for Aging

Everything we know is tending toward chaos (unless there is outside inter-vention), towards an equilibrium with the environment. Von Bertalanffy wrote that the significance of the second law can be expressed in another way. It states that the general trend of events is directed toward states of maximum disorder, the higher forms of energy such as mechanical, chemical, and light energy being irreversibly degraded to heat, and heat gradients continually disappearing. Therefore, the universe approaches entropy death when all energy is converted into heat of low temperature, and the world process comes to an end.

It was not until the 1950s that entropy started to seriously emerge in discussions of living phenomena. Complicated biological processes such as cell differentiation, growth, aging, etc., were now analyzed from the second law of thermodynamics and entropy calculations made. Bailey wrote that entropy is a very viable concept for the biological and social sciences. It applies to both open and closed systems. It can be discussed in terms of orga-nization or order. Jones said that one common feature of biological processes is their unidirectionality in time, that is, they are not reversible, except under special circumstances. Since entropy is the only physical variable in nature which generally seems to parallel the direction and irreversibility of time, these should be fertile areas for the effective use of entropic models.

Zotin proposed that we evolve toward a final state, death, by a series of changes, each change called a stationary state. We settle into a stationary state, stay for a while, until pushed to the next, and the next. This is clearly seen in the transformation of butterfly larvae and pupae, dramatic physical changes. Balmer applied entropy concepts to the study of an aging annual fish. This species displays all the characteristics of birth, growth, aging, and senile death, over a short twelve-month period.

If death represents the ultimate disorder (and our maximum entropy content) then we can characterize our longevity potential and vitality by the distance we are from the death. We can calculate a difference in entropy content, from the present (where we are) to where we're heading (towards death). The difference is called Excess Entropy (EE) and is a driving force for life. By tracking EE versus age, we can see our life unfolding, or winding down, as EE approaches zero, the end of the journey. Not only is Excess Entropy (EE) an important parameter in tracing our lifespan, but so too is the rate of change of Excess Entropy. In other words, how quickly EE is diminishing with age is another key marker. We call this second parameter Excess Entropy Production (EEP).

EEP is the rate of change of EE with time, dEE/dt. From Prigogine's theory of minimum entropy production, we can surmise that EEP should not only become a minimum in the vicinity of death but EEP should indeed become zero since that final stationary state is also the final equilibrium state where all thermodynamic forces and flows become zero. It is calculated from the Prigogine–Lee–Hershey formulation, which shows EEP being proportional to the rate of change of a thermodynamic force (chemical affinity) and flow (chemical reaction velocity). Using the oxidation of fats, proteins and carbohydrates as the basic chemistry, chemical affinity and chemical reaction velocity expressions are developed and used to calculate EEP: (Daily Protein Consumption minus Minimum Protein Requirement)^2 divided by Daily Protein Consumption. After the EEP versus age lines are drawn, EE is obtained from $EE = \int EEP\, dt$. Thus, we trace life's course by tracks given by EE and EEP. The Lee–Hershey theory developed from some of Prigogine's ideas predicts the behavior of EE and EEP. Essentially, Lee and Hershey postulate that the EEP line should diminish steadily, and end in the vicinity of the final stationary state (equilibrium or death), where EEP becomes zero. Also, EE should tend towards zero as our entropy content approaches its maximum at death. With time, as we age, our entropy content approaches the maximum, and so EE tends towards zero.

Longevity Predictions Based on EEP and EE Tracks

Using male pooled data the EEP track declines steadily with age, ending at zero when the age of about 85 has been attained. For these pooled data, a lifespan track projects an expected lifespan of about 85 years for the average male.

Summarizing the EEP and EE Theory of Aging

The Excess Entropy Production (EEP) and Excess Entropy (EE) theory of longevity tracking yields the following.

1. EEP is positive and decreasing (type I)
2. A stress period can be identified, it being a time of sustained, increasing EEP (type II). A killing stress is one where death occurs while in the stress period. A survivable stress is defined by the resumption of a normal, decreasing EEP history after the stress period (type III).
3. Death is impeding when the EEP track approaches zero.
4. Excess Entropy (EE) is negative.
5. EE ascends to zero in the vicinity of death.

Life and Death

Aging can be considered a continuous process. Some have speculated that the aging process starts immediately after the fertilization of the egg. Aging may start following the attainment of adulthood, after the periods of childhood and adolescence, which are times marked by rapid growth and major body changes. Beyond these phases of life the BMR and EEP steadily decrease with age; the body is slowing down.

Aging may be the evolution towards a more probable state, the equilibrium state. Schrodinger wrote that living systems survive by avoiding the rapid decay into the inert state of equilibrium. Ritenberg and Jones proposed that the approach to equilibrium is a sign of death. Death may also be thought of as the attaining of a critical, maximum state of entropy during our journey towards equilibrium. Schrodinger further stated that a living organism continually increases its entropy and tends to approach the dangerous state of maximum entropy, which is death.

Death can be when a critical amount of randomness is attained, when a certain amount of disorganization is suffered. Thus aging is a randomizing process, a disorganizing process. In terms of stability theory, equilibrium is the point or region of attraction; we are drawn relentlessly towards equilibrium. Death during type II behavior can be envisioned as a catastrophic event since it is an abrupt transition (life to death) that tends to occur at an age less than the expected. Life may be considered analogous to the spring-wound watch, where the timepiece may stop by one of two possible mechanisms. It can simply wind down (approach equilibrium). This is the watch's version of type I death. Or the internal mechanism can somehow fail and the watch "dies" prematurely. This is equivalent to type II death. (Since this is not a self-repairing system, a type III track is probably not possible.) Thus the type I and III deaths can be described as equilibrium deaths, and type II as instability or catastrophic death.

Life may be considered a temporary upset from or perturbation of equilibrium. Equilibrium is absolutely stable, a universal attractor. Equilibrium always wins. Aging is a spontaneous process, where the body dissipates its Excess Entropy (EE). The various theories of aging (cross-linking, wear and tear, free radical, etc.) all imply a declining metabolic rate with age, and by extension, a diminishing Excess Entropy and Excess Entropy Production. Evolution may be the natural process of prolonging the time that an organism spends in the far-from-equilibrium state. Genetics still plays its role in determining longevity potential but is not in conflict with the ideas presented here. The tendency to return to equilibrium will always be expressed; the entropy laws will always apply. Death will always have a probability of absolute certainty.

CHAPTER 9

Excess Entropy (EE) and Excess Entropy Production (EEP)

Internal entropy production for a chemical reaction system is given by Prigogine and Wiame

$$dS/dt = Ar/T, \tag{1}$$

where S = internal entropy content; A = chemical affinity, a chemical driving force; r = reaction velocity, a chemical flow; T = temperature; t = time.

Equation (1) can be more generally written as

$$\sigma(S) = dS/dt = \sum_{j=1}^{n} J_j X_j, \tag{2}$$

where $\sigma(S)$ = internal entropy production at any time, t; J_j = a thermodynamic flow, for component, j; X_j = a thermodynamic driving force, for component, j.

For a reference state, from Eq. (2), we can write

$$\sigma^0(S) = \sum_{j=1}^{n} J_j^\circ X_j^\circ \tag{3}$$

and

$$\sigma(S) = \sigma^0(S) + \delta\sigma(S), \tag{4}$$

where $\delta\sigma(S)$ = a small deviation of internal entropy production from the reference state.

If Eq. (4) is rearranged, the result is

$$\delta\sigma(S) = \sigma(S) - \sigma^0(S), \tag{5}$$

or, from Eqs. (2) and (5),

$$\delta \left[\sum_{j=1}^{n} J_j X_j \right] = \sum_{j=1}^{n} J_j X_j - \sum_{j=1}^{n} J_j^o X_j^o. \tag{6}$$

It was shown by Prigogine, that $\sigma^0(S)$ in Eq. (3) is a minimum in the reference state (the equilibrium or stationary state) so that the left hand side of Eq. (6) is always positive. From Eq. (2), Lee wrote

$$\delta(dS/dt) = \sum_{j=1}^{n} J_j \delta X_j + \sum_{j=1}^{n} X_j \delta J_j \tag{7}$$

and using the commutative property

$$\delta(dS/dt) = d(\delta S)/dt,$$

Equation (8) can be obtained

$$d(\delta^2 S)/dt = 2 \sum_{j=1}^{n} \delta X_j \delta J_j, \tag{8}$$

where

$$\delta^2 X_j \ll \delta X_j$$
$$\delta^2 J_j \ll \delta J_j,$$

and $\delta J_j = J_j - J_j^o$, the deviation of the flow from the reference state; $\delta X_j = X_j - X_j^o$, the deviation of the force from the reference state.

By expanding entropy, S, in a Taylor series about the reference state, we can obtain,

$$S = S^0 + \delta S + 1/2\delta^2 S + \cdots, \tag{9}$$

where S^0 = entropy of the system, in the reference state; δS = first entropy deviation from the reference state; $\delta^2 S$ = second entropy deviation from the reference state; and $\delta S = 0$, if the reference state is an equilibrium or stationary state of maximum entropy. Therefore, from Eq. (9) and $\delta S = 0$ we can obtain

$$S - S^0 = \frac{1}{2}\delta^2 S = \text{Excess Entropy}. \tag{10}$$

It is this difference between the entropy content, S, at any time and that of the reference state, S^0, which is defined as Excess Entropy, i.e., $EE = S - S^0$. If the reference state is an equilibrium or stationary state, S^0 is a maximum. Thus Excess Entropy (EE) is always negative and approaches zero in the negative domain as the system ages, and evolves towards the reference state. Excess Entropy can be considered a thermodynamic driving force for the life process.

From Eqs. (8) and (10), we can define Excess Entropy Production (EEP) as

$$EEP = dEE/dt = d(\delta^2 S)/dt = 2\sum_{j=1}^{n} \delta X_j \delta J_j. \tag{11}$$

It was demonstrated by Hershey and Lee that EEP approaches a minimum or zero as the system approaches an equilibrium or stationary state. EEP describes the rate of approach of EE to the final state.

For a chemically reacting system

$$Z + Y \rightarrow C + D, \tag{12}$$

and with the definitions A = chemical affinity = $\log[ZY/CD]$ = chemical force, X; and r = chemical reaction velocity = ZY = chemical flow, J, we can obtain Eq. (13), using Eq. (11) and assuming Y, C, D are constants

$$EEP \cong (\delta Z)^2/Z. \tag{13}$$

Although the influences of free radicals, vitamins, minerals and other nutrients are essential in establishing longevity, nevertheless in a living system the chemistry of life is basically the metabolism of carbohydrates, fats and protein. In general, in a homogenous population, the proportions of these food components in the diet tend to remain approximately fixed. Thus, it is useful to focus on only one of these, protein, for example. Equation (12) now becomes:

protein + oxygen $\rightarrow$ carbon dioxide + water + urea + energy, and Eq. (13) is as follows:

$$EEP \cong (\delta[\text{Protein}])^2/[\text{Protein}], \tag{14}$$

where [Protein] = daily protein consumption; δ[Protein] = daily protein consumption minus the minimum protein required (in the reference, equilibrium or stationary state of maximum disorder).

The minimum protein required was chosen as that ingested to sustain life, obtained from data on elderly persons in the vicinity of maximum disorder or death. This is approximately 56 grams per day for males, 48 for females.

Hershey and Lee have explained how Excess Entropy and Excess Entropy Production data can be used to develop longevity tracks and make lifespan projections for the individual. It is also possible to use these tracks as a diagnostic tool to spot incipient illness before they are clinically manifested.

In theory EEP goes to zero as we approach "senile" death, which is the final, stationary state, the equilibrium or death condition. Hershey, *et al.* demonstrated this by a graph of EEP versus Age, for males and females, with data obtained from the NIH. Where the lines intersect the Age axis at EEP = 0 is the theoretical, projected death age for males and females, about 84 and 100 years respectively.

Comparison of Theoretical Longevity Projections versus Actual Death Age for Individual Subjects

Dead People Project

Comparison of Excess Entropy Production (EEP) and Excess Entropy (EE) Theoretical Longevity Projections versus Actual Ages of Death for Individual Deceased Subjects

---

Theoretical longevity projections can be made based on the thermodynamics concepts, Excess Entropy Production (EEP) and Excess Entropy (EE), which are derived from non-equilibrium thermodynamics principles developed by Ilya Prigogine (Nobel Prize winner, 1977) and Daniel Hershey. The theoretical death age is when the EEP or EE numbers go to zero.

The accuracy of this method is expressed in values of K, where

$$K = \text{Actual Death Age}/\text{EEP or EE Theoretical Death Age}$$

Part I

We used NIH Basal Metabolic Rate (BMR) data from the files of 39 deceased subjects, and our own theoretical equations, to calculate the EEP or EE history for each subject. We found $K = 0.98$, an average value covering all 39 subjects. This suggests the theoretical longevity projections may be able to predict the actual death age for the individual to within 2%.

Part II: Maple Knoll Retirement Village Data

We gathered blood pressure and heart rate histories for 15 deceased subjects who resided at Maple Knoll. From these data and our correlation which relates these blood pressure and heart rate histories to the BMR, we again calculated EEP and EE trendiness for the 15 individuals (11 EEP and 11 EE tracks). Again we found EEP and EE theoretical death ages. For the Maple Knoll subjects, $K = 1.00$, an average of the 22 histories (15 subjects).

Summary of the Results from Parts I and II	
Part I	*Part II*
NIH Data	Maple Knoll Data
EEP Histories	*EEP Histories*
31 Subjects (Types I + III)	11 Tracks
$K = 0.97$	$K = 0.96$
Std. Dev. $= 0.06$	Std. Dev. $= 0.03$
EE Histories	*EE Histories*
8 Subjects (Type II)	11 Tracks
$K = 1.03$	$K = 1.05$
Std. Dev. $= 0.06$	Std. Dev. $= 0.05$
EEP and EE Histories	*EEP and EE Histories*
39 Subjects (31 + 8)	22 Tracks (11 + 11)
$K = 0.98$	$K = 1.00$
Std. Dev. $= 0.07$	Std. Dev. $= 0.06$

Consequences of the Results of Parts I and II

This EEP/EE technique can be useful as a medical diagnostic tool. For living subjects, by producing a continuous record of EEP and EE versus age, we can easily see any significant changes in the nature of the trendlines, and changes in the theoretical longevity projections. For example, incipient or early stage cancer should change the trendlines significantly.

PART V

Entropy Theory of Aging Systems: The Corporation

CHAPTER 10

Evolving Corporate Systems: Entropy, Aging and Death

The change process for an open system, influenced by the size and complexity of its structure, may be driven by its entropic distance from equilibrium. It evolves by going through a series of non-equilibrium, dissipative states. Life may be a non-equilibrium, entropy-driven process. There is increasing disorder with age. Death is a state of maximum entropy. These ideas are applicable for so-called inanimate corporate systems and civilizations.

Size versus Function

A city or nation becomes too big when it can no longer provide its citizens with the services they expect — defense, roads, health, courts, etc. — without amassing such complicated institutions and bureaucracies that they actually end up preventing the very goals they are attempting to achieve. Social problems expand at a geometric rate with the growth of a city or nation while the ability to cope with them expands only arithmetically.

The problem then may be how to stop growing: division when size becomes overbearing. Beyond relatively narrow limits, additional growth detracts from efficiency and productivity. In evaluating the critical size of a society, it is not sufficient to think only in terms of the size of its population. Density and the velocity of information flow (related to the extent of its administrative integration and technological progress) must likewise be taken into account.

Larger — excessively large — communities require saturation police forces to match at all times the latent power of the community. The answer to the problem may reside not in increased police power but in a reduction of social size — the dismemberment of those units that have become too

133

big. Below a certain size, everything fuses, joins or accumulates; beyond the optimal size, everything divides collapses or explodes. The stars in the sky sometimes grow to the point where, instead of generating energy, they begin to absorb it (as great powers do in the political universe). Being too small leads to a self-regulating device through aggregations or fusions until a proper and stable size is achieved and the functions determined or teleological form is fulfilled.

For the aging system, what was previously flexible and swift now is slow and rigid, with an accumulation of turgid or inert bulk. In an overaged social system, a powerful authority is required to move its obdurate cells. If opportunism deteriorates as a result of overgrown cells, then it follows that we may restore proper function through cell division and the reintroduction of a small-cell arrangement. Division represents a cure principle and progress, while unification seems to portend disease.

If large bodies are inherently unstable in the physical universe, they are in all likelihood unstable also in the social universe. What is applicable in the universe as a whole and in special fields such as biology, engineering and art should also be appropriate in politics. Small states with their concise dimensions and relatively insignificant problems of communal living give their citizens time and leisure; large powers with their enormous social demands consume practically all the available energy of their servants and citizens in the mere task of keeping their immobile, clumsy society functioning.

The larger more powerful societies find more of their products devoured by the task of coping with the murrain of its power. So they build the bombs and tanks and traffic lights, parking lots, etc. Business cycles, attributed mostly to capitalistic countries, are nevertheless phenomena also of socialistic systems; the problem is not the ideology. Once beyond a certain size, it becomes increasingly difficult to deal with economic perturbations, the cycle of boom and bust. The overshoot and undershoot of the cycles seem to become more closely linked to each preceding cycle, and we tend towards instability — seemingly unable to dampen the fluctuations.

The performance of a corporation after it has reached a certain size begins to decline in spite of the illusionary fact that the total output may continue to rise. Wise businessmen will not extend production to maximum capacity but to optimum capacity. Instead we should build new and independent plants and begin the battle against diminishing productivity again, but now with a small cell model. Successful social organisms, be they empires, states, countries, cities or corporations seem to have in all their diversity of language, customs, traditions and governing systems one common feature: the small-cell pattern.

Emerging States Driven by Non-equilibrium Conditions

The behavior of corporations, cities, civilizations, living cells, economic processes, ecological systems and even transportation networks illustrate the recondite dynamics of non-equilibrium open systems. These systems near equilibrium can be buffeted by small perturbations in energy and mass pressures; but no new organizations, no new structures are formed. Imposing stronger gradients from the outside world forces the appearance of new, dissipative, non-equilibrium states. On the other hand economists have a blind faith in reversibility (we can restore original conditions by invoking the same laws, backwards and forwards and ignoring time). Until now, they have stressed economic balances of energy and matter — what comes in must equal what goes out — rather than an understanding that something is lost in every transaction (in entropy terms), in transforming raw materials or energy. The real world dictates the transformation in one direction only: low entropy to high entropy.

Aging Civilizations

What is it that drives us to new structures? Is it creativity, challenge and response, the stimulus of different environments (terrain), the stimulus of penalization such as religious discrimination? A society continues to grow, it seems, when successful response to a challenge provokes a fresh challenge to be met, converting a single action into a series of movements. The movement of challenge-and-response becomes a self-sustaining series if each successful response provokes disequilibrium, requiring new creative adjustment. Growth is assured through integration to differentiation to reintegration and hence to redifferentiation. Civilizations have met their death not from the assault of external and uncontrollable forces but by their own hands. We become complacent and allow custom and mimesis to become entrenched. The forces acting upon the arrested society demand so much energy simply to maintain the position already attained; there is nothing left for reconnoitering the course ahead.

A society in the process of dissolution is usually overrun and finally liquidated by so-called barbarians from beyond its frontiers, the pressure built up by frontiers which have become seemingly impermeable. The fluid zone of contact between the state and its neighbors has become frozen into a seemingly impenetrable military frontier: a closed system where neighbors become enemies and cultural exchanges cease. Hadrian's Wall was visible evidence of the Roman Empire's attempt to insulate itself from the barbarian tribes of northern Britain and to protect against invasion.

Entropy, Aging and Death

We can say that entropy is a measure of the disorder of a system and show that more disorder means higher entropy content. Real processes tend to go in the direction of increasing entropy. Aging can be envisioned as an irreversible process of entropy accumulation. Getting older means having less control of body functions, being more disordered.

Death is the ultimate disorder, a state of maximum entropy. The second law of thermodynamics essentially says that systems will run down of their own volition if left to themselves. In other words, the entropy content tends towards a maximum. Thus increasing entropy could be an indication of the direction in which the system is inclined to go. Unless there is outside intervention, the second law of thermodynamics codifies the one-sidedness of time, or time's arrow. We can only move forward, that is, time is irreversible. Everything we know is tending towards chaos (unless there is outside intervention), towards an equilibrium with the environment.

Much of the historical development of entropy has dealt with isolated or closed systems. The second law of thermodynamics states that a closed system must evolve to a state in equilibrium with its environment — a condition of maximum entropy. Open systems are those which can exchange both matter and energy with the surroundings. Obviously we, the living, are examples of open systems. Open systems must maintain the exchange of energy and matter in order to sustain themselves, or slow the approach to the final state, death. We seem to evolve towards a final state, death, by a series of changes, each change called a stationary state. We settle into a stationary state, stay for a while, until pushed to the next, and next.

Equilibrium Versus the Stationary State

In describing the aging process, we need to distinguish equilibrium from the stationary (steady) state. Suppose I've kept a piece of iron in my refrigerator for a very long time. We can assume that the temperature of the iron is close to that of the refrigerator interior. Now I remove the metal from the refrigerator and place it on a table in my kitchen. We know that the metal will begin to warm, heat flowing into the iron from the surrounding, warmer air. The driving force for heat flow is the temperature difference, between the ambient air and the iron. In the beginning, with the greatest temperature driving force, a relatively large amount of heat enters the iron, warming it quickly. The warming iron receives lesser and lesser quantities of heat, since the temperature driving force between the metal and the air is diminishing. After a few hours, the temperature of iron is almost that of the room, the temperature driving force is approaching zero and the heat flow to the iron is negligibly small. If we wait long enough, we can say that the iron is now in equilibrium with its environment; the temperature driving force is approaching zero, as is the heat flow. Now nothing is changing as there are no driving forces to cause change.

On the other hand, suppose I've got a tank half filled with water. There is a pipe attached which will allow new water to enter the tank and we have a drain at the bottom through which water exits. If the amount of water entering is constant and equal to the water exiting, then the liquid level in the tank remains fixed. To the observer, things look steady. Though there are driving forces available, these forces are balanced and nothing changes. We say we are at a stationary (steady) state. (This is not to be confused with equilibrium, where the driving forces are zero.) I could generate a new stationary state in the tank by changing the incoming flow of water and noting the new level of water in the tank when things stabilize.

Aging as an Entropy Driven Process

We can consider the aging process as a series of steps, proceeding from one stationary state to the next and the next. We can talk of an entropy driving force, causing the transition from one stationary state to the next. Each time we achieve a stationary state, we "rest" for a "moment" and then go on to the next. In each stationary state, we hunker down, collect ourselves, minimize our entropy production, smooth out our chemistry, and await the next push. And so we age. Thus the stationary state can be characterized as an entropy (disorder) producing state, where this entropy production is drawn down to a minimum, before going on to the next stationary state where a new minimum is established. The theorem of minimum entropy production is a fundamental concept of the stationary state. Death is the final state to which we are drawn, where there no longer exists a tension for life; the entropy driving forces have been reduced to nothing or some minimum level, below which life cannot be supported.

Aging may be the evolution towards a more probable state, the equilibrium state. As the body ages (returns to equilibrium), driving forces weaken. Thus a living organism continually increases its entropy and tends to approach the dangerous state of maximum entropy, which is death. Death can occur when a critical amount of randomness is attained, when a certain amount of disorganization is suffered. Thus aging is a randomizing process, a disorganization process.

Life may be considered analogous to the spring-wound watch, where the timepiece may stop by one or two possible mechanisms. It can simply wind down (approach equilibrium). Or the internal mechanism can somehow fail and the watch "dies" prematurely. Life may be considered a temporary upset from or perturbation of equilibrium.

The Change Process in Corporate Systems

Enlightened corporate leaders see change as necessary for the orderly transition — the gradual evolution — from one stable state to another. They propose a departure from equilibrium, and enter the arcane world of non-equilibrium (dissipative) states. Instead of seeking the security of constancy, these proponents of non-equilibrium dynamics say that organizations proceed to and through various plateaus, resting on each level until displaced by the convergent forces of energy, material, and information exchange with the environment.

These open systems are driven from one seemingly stable orbit of operations into another. The plan is to remain viable by switching to a new dynamic regime, yielding, in a sense, order through fluctuations, which is the reverse of the behavior of some systems near equilibrium. The organization near equilibrium attempts to meet new pressures for change by damping them out and returning to its original conditions after these small, temporary deviations. Some might claim the forces of racial integration were like this, where for a while, the integrationists' pressures were met by sufficient resistance to continue the original segregated condition. There seemed to be stability in the segregated state but this was specious since the time frame was not sufficiently long to see the trends. Actually, the segregated state was inherently unstable. When the forces for integration became morally irresistible, the system moved to a new non-equilibrium position, which is the present point in history. Thus instead of having a new civil war over the issue, society found order through fluctuation or change. In this new state our response to old pressure is met with more degrees of freedom. There are more ways to meet old problems. With the new dynamics of order through fluctuations, the challenge is to delineate the bounds of stability and to attempt to identify the new state which we wish to go to. There can be surprises along the way, of course. For example, a system can be driven too far in size and complexity, as some think New York City has been. The high density of the population and the limited system of roads in Manhattan has led to the surprising result that during rush hours it is often quicker to walk in Manhattan than it is to drive the same distance in an automobile.

These dissipative, non-equilibrium states might in some simple cases be described with appropriate feedback relationships. If the feedback is negative, this feedback or return of knowledge and information to the pressure point is considered a control on the pressure fluctuation. The tendency is toward stability and a subsiding of the system to its original state. Should the feedback be positive, we have a condition where there is reinforcement of the amplitude of the fluctuations. (We add a positive pressure to a positive feedback.) This yields a higher output, increasing the positive feedback value, which when added to the positive pressure gives a still larger output. Presumably a point could be reached where the sheer size of the fluctuation drives the system to a new non-equilibrium or dissipative structure.

Whether considering organizations such as corporations, societal structures such as the welfare system, human living organisms, or the universe, what emerges is a general scheme for change, a way of going from one non-equilibrium state to the next non-equilibrium state: order through fluctuations. The change process is accomplished by deviation amplifying or positive feedback means. This might be called a revitalization process, requiring explicit intent by members of society. To get there from here, the system needs to go through these steps: (1) achieve a non-equilibrium plateau; (2) experience stress; (3) endure cultural distortion; (4) plan for revitalization; and (5) enter a new non-equilibrium plateau.

Social theories have traditionally been geared to structure, not process, and to ideals of equilibrium and structural stability. The emphasis has been on steady state and negative feedback (which corrects deviations). It may be that the analysis and planning of social organizations now may need to be brought into consonance with the newer non-equilibrium order through fluctuations approach. Contemporary society is characterized by a rapid dissemination of information. In the past, science and technology were the chief agents of positive feedback (deviations from existing norms were amplified) in transforming the human-environment relationships. On the other hand, societal philosophy and techniques provided negative feedback to stabilize the relationships. But the deviation-amplifying process can increase differentiation, develop structure, and generate complexity. This may work to enrich society and allow us to move on to something better than we had previously. Much of the unpredictability of history is perhaps attributable to deviation amplifying causes. The result is either a runaway situation or evolution.

Entropy Concepts Applied to a Corporation

Entropy is a measure of the disorder in a system. It increases as differences within the system are dissipated and the number of organizational units is raised. Entropy tends towards a maximum in the vicinity of death for a living system as control is lost. The corporation tends towards maximum entropy when all its units are independent and equal (a disaster or chaos condition). Entropy decreases to a minimum, ideal value when all corporate units have perfect access to the leader and to each other. Tendencies towards verticality in a organization structure increase entropy and drive efficiency down.

Excess Entropy is a measure of the entropic distance from disaster. It approaches zero as the living system as well as the corporate system approach disaster or chaos or death; Excess Entropy achieves its greatest values when the corporation moves towards an ideal structure.

Excess Entropy Production is the rate of change with time of Excess Entropy. For living systems, Excess Entropy Production diminishes with age and nears zero in the vicinity of death. For the aging corporation, a diminishing Excess Entropy Production track can signal stagnation, a general decline in organizational vitality.

The driving force for change, the motivation factor which drives a system through its non-equilibrium stationary states can be Excess Entropy. It expresses the tension of life, the distance (in entropy terms) from equilibrium (death, disaster, chaos). Excess Entropy Production measures the speed of approach to equilibrium.

CHAPTER 11

Entropy and Size, Structure, Stability, Senescence

Calculations of entropy changes in open systems were reported in 1896, entropy production in 1911. Clausius coined the word, entropy, from the Greek language to mean transformation and defined it as the increment of energy added to a body as heat divided by its temperature, during a reversible process. Brillouin showed a connection between entropy and information. Dowds developed his own method of applying entropy concepts to the examination of the data from oil fields, as did others in land use planning, and in Nigeria to analyze water runoff. It was the work of Shannon that more firmly connected information and entropy. Suppose entropy measures not the information of the sender but the ignorance of the receiver, removed by the receipt of the message. The probabilities assigned individual messages are a means of describing a state of knowledge. Historians such as Toynbee, Spengler and Spencer believed in historical determinism, associating civilization with the living tendencies of birth, maturation, senility and death.

Shannon's formula connects informational entropy to the probability that the system is in a certain structural configuration. A number from one to a hundred can always be identified in seven guesses (which can be answered by "yes" or "no"). To do this one needs to make guesses in such a way as to eliminate one-half of the remaining range. By applying Shannon's formula we can calculate the informational entropy for this process; we get the number of guesses required to find the unknown number, or the amount of entropy information associated with the process. The relationship between entropy and information provides further clarification of the fundamental principle of the living process: increases in entropy can be viewed as a destruction of information. A living, open organism must be able to maintain its organized state against the forces of disorganization; it must continue homeostasis and a purposeful behavior.

Disturbed slightly, we will return to the same steady state conditions as soon as the tamping ceases. Body temperature is an example of such behavior; with the invasion of foreign organisms, the antibody-antigen reaction results in a rising of the temperature set point and we develop a fever. When the external influences diminish, we return to our normal temperature. If however small departures from steady state are magnified so that the system moves even further away, the steady state is unstable. There may be several steady states or plateaus, as suggested by Prigogine: stability through fluctuations. The more the system is organized (in our youth), the more it is equipped to resist killing stresses, a situation reversed in old age when there is more disorganization and less ability to resist disturbances which overcome our stability criteria. In the theory of social systems, many analogies suggest themselves; the institution can be easily imagined to be an organism. Its organizational structure can be thought to correspond to our anatomy, its modus operandi to physiology, its history to our development and evolution. Mathematical models of power relations among states have been drawn to help clarify the distinction between stable and unstable states. If states in concert and vying with each other for power constitute a system, then the system also has certain properties of stability. Economic systems are also included here; to the extent that certain aspects of an economic system (fluctuations in production levels, prices, or investment capital) can be cast in a mathematical model, questions about equilibrium, stability and instability can be answered by mathematical deduction rather than by intuitive guesses. The models are based on the assumption that there can be transitions of a system from state to state, governed somewhat by probabilities. In a large population, probabilities become frequency of occurrence.

In the matter of nation systems, size may govern stability. The notion that size governs is one that has long been familiar; Haldane showed many years ago that if a mouse were to be as large as an elephant, it would have to become an elephant, that is, it would have to develop those features such as heavy, stubby legs to allow it to support its weight. City planners realize that accumulations of people much above 100,000 create entirely new problems: it is virtually impossible for a city exceeding this size to run in the black since the municipal services it must supply cost more than any feasible amount of taxation can raises. Social problems expand at a geometric ratio with the growth of a city while the ability to cope with them expands only arithmetically. Nuclear explosions result when a certain critical mass is reached. Cancer represents a group of cells outgrowing their

normal bounds. Isn't it true that human beings, charming in small aggregations, become mobs when over-concentrated. The problem then may
be how to stop growing: division when size becomes overbearing. Arnold
Toynbee, linking the downfall of civilizations not to fight amongst nations
but to the rise of universal (large) states, suggested that we return to the
Greek ideal of a self-regulatory balance of small units. Kathleen Freeman
in a study of Greek City-States showed that nearly all Western culture
is the product of the disunited small states of ancient Greece and that
these same states produced almost nothing after they became united under
Rome. Justice Brandeis devoted a lifetime to exposing bigness by demonstrating that beyond relatively narrow limits, additional growth of plant
or organizational size no longer adds to, but detracts from the efficiency
and productivity of firms. Henry Simmons asserted that the obstacles to
world peace do not lie in the alleged anachronisms of little states but in the
great powers and suggested their dismantlement. Kohr claimed the prin-
cipal immediate cause behind both the regularly recurring outbursts of mass
criminality and the accompanying moral numbness does not seem to lie in
a perverted leadership or corrupt philosophy but is linked with frequency
and numbers. In a small society, the critical quantity of power can only
rarely accumulate since the cohesive force of the group is easily immobilized by self-balancing centrifugal trends of individuals. In evaluating the
critical size of a society, it is not sufficient to think only in terms of the
size of its population. Density and the velocity of information flow (related
to the extent of its administrative integration and technological progress)
must likewise be taken into account. A large population thinly spread may
act as a small society. Similarly, a volatile and faster moving society may
transgress the bounds into an unstable state: an agitated crowd in a theater
may tax the heretofore adequate number of exits. Larger — excessively
large — communities require saturation police forces to match at all times
the latent power of the community. The numbers are simple to handle in
small social units. The answer to the problem may reside not in increased
police power but in a reduction of social size — the dismemberment of those
units of society that have become too big. Aggression, in countries as well
as communities, arises spontaneously, irrespective of nationality or dispo-
sition; the moment the power becomes so great that in the estimation of its
leaders the system has outgrown the power of its prospective adversaries.

The stars in the sky may seem huge, but what are they in relation to
space itself? They sometimes grow to the point where instead of generating
energy, they begin to absorb it (as great powers do in the political universe).

The effort to maintain their existence forces these stars to consume more than they receive, living off their capital until the supply of hydrogen becomes exhausted. The stars collapse and in the process of collapsing generate rotary forces comparable with gravity, causing fantastic explosions as they disintegrate: the supernova. There is instability of the microcosm also. Being too small leads to a self-regulating device through aggregations or fusions until a proper and stable size is achieved. Humans aggregations must have magnitudes drawn from our inherent stature, and be measured in miles and years, not parsecs and eternities. For us, disease and aging produce an upset in our rhythm of life. What was previously flexible and swift now is slow and rigid. Big-power systems also move and live, though like an old person, at a reduced speed, with the accumulation of turgid or inert bulk. A good balance of systems in the world, be it of stars, states or people must be flexible and self-regulatory, derived from the independent existence of a great number of small-component parts held together not in tight unity but elastic harmony. If opportunism, necessary to living systems, deteriorates as a result of overgrown cells, then it follows that we may restore proper function through cell division and the reintroduction of a small-cell arrangement. Language conveys information through a division of sounds. Parties may be saved from boredom not by having all guests assembled in a single circle dominated by a magnetic personality, but by dissolving the pattern of unity into a number of small groups. By branching off into a number of different forms, orders, classes and subclasses, an originally unified group diversifies itself. The first step towards a higher form of life was accomplished when living substances differentiated into green plants, bacteria, fungi and animals. If large bodies are inherently unstable in the physical universe, they are in all likelihood unstable also in the social universe. What is applicable in the universe as a whole and in special fields such as biology, engineering and art should also be appropriate in politics. The law of crowded living is in other words, organization; the greater the aggregation, the more dwarfish we become. But what is the ideal size of a state? Up to what point can a political community grow without endangering the existence of the state and the individual? Or how small can it become before we achieve a similar denouement? When does a community become stable? And for how long? Small states with their concise dimensions and relatively insignificant problems of communal living give their citizens time and leisure, without which no great art can develop; large powers with their enormous social demands consume practically all the available energy of their servants and citizens, in the mere task of keeping their immobile, clumsy society functioning. Forever afraid of cracking beneath their own oppressive weight, gigantic states can never release their population from servitude to their collective enterprise. They are deflected from the grace of individual living to the puritan virtue of cooperation, which is the law of some highly efficient animal societies. Arnold Toynbee suggested a consanguity between cultural productivity and relief from exacting social tasks in nations, states and churches. What is great in great nations is not to be found during their periods of power (which kept them busy with occupying the limelight on the stage of history) but during the time when they were relatively insignificant and little. In the large state we are forced to live in tightly specialized compartments where life's experiences are carefully circumscribed, whose borders we almost never cross. According to Kohr, the great empires of antiquity, including the Roman Empire have not created a fraction of the culture which the ever-feuding Greek city-states produced. The great empire's chief accomplishments were technical and social, not cultural.

This behavior of nations and other human organizations illustrates the recondite dynamics of the non-equilibrium open system, a real world application of the theory. States, living cells, economic processes, and ecological systems, transportation networks: we can create the bridges between the physical and social sciences. Prigogine attempted to draw consistent mathematical analogies between chemical systems, entropy and social processes; Georgescu-Roegen did the same for economics. Prigogine studied self-organization under conditions of fluctuations and change, an evolutionary process in systems pushed far from equilibrium. Whether the fluctuations impinge on a household or a nation, the belief is that they can be explained. He examined oscillating phenomena and sought a basis for predicting order from the perspective of entropy production (regarded by some as the reification of the arrow of time since it describes in general the direction spontaneous process must go). Systems near equilibrium can be buffeted by small perturbations in energy and mass pressures but no new organizations, no new structures are formed. Imposing stronger gradients from the outside world could force the appearance of new, dissipative, non-equilibrium states. Examples cited by Prigogine are a town and the living cell. Georgescu-Roegen wants economics, which ignores entropy, to begin to mark the existence and applicability of non-equilibrium dynamics and irreversibility. Economists, he says, have a blind faith in reversibility (we can restore original conditions by invoking the same laws, backwards and forwards and ignoring time). Matter is subjected to entropy degradation, from higher order to lesser, and hence becomes less useful. What Georgescu-Roegen is stressing is an evolutionary philosophy. If driven hard enough we see new structures developing, nurtured by energy and matter fluxes. Entropy, applied to the economic process, adds important new perceptions to the interactions of humans, technology, the market system and limited resources. Ideas spawned by Georgescu-Roegen. He proposed we think in terms of irreversibility, limits on resource availability and a more parsimonious society. Most current economic policy tinkers with prices, taxes or the market in some way. Until now, we have stressed economic balances of energy and matter — what comes in must equal what goes out — rather than an understanding that something is lost in every transaction (in entropy terms), in transforming raw materials or energy. The real world dictates the transformation in one direction only: low entropy to high entropy. The consumer takes in high-grade, ordered energy and matter and exhausts low-grade, disordered wastes. The wastes must not injure or render inoperative the feedback and control mechanisms which affect the stability of the open, temporary state. Consumers may be individuals, cities, governments, corporations, civilizations. The entropy-economics link can be applied to feasibility studies in the recovery of old oil wells, accounting and cash flow, agriculture, new product development, decision-making and information flow. Money constitutes the economic equivalent of low entropy.

Biological systems maintain a state of high coherence, which is perhaps the most striking feature of living things. They may evolve to new, temporarily organized states called dissipative structures, to signify that they are created and maintained by the dissipative, entropy-producing processes within the system: regulatory processes at the cellular level. Viewed as a succession of instabilities within a fundamentally irreversible, self-organizing domain, the changes are achieved spontaneously, according to their own time or age scale. Some believe evolution proceeds this way; each step is followed by another which has a greater chance to occur (and hence takes less time to appear). Autocatalysis. In order to describe the behavior of a self-organizing system we need some index of the intensity of energy dissipation which will provide us with a clue as to its organizing proclivity.

Describing a self-organizing process requires us to define and quantify the notion of order; we need a model to account for the transition from one slot in the hierarchy to the next. The degree of organization of a biological system is established teleologically (with respect to its purpose): seeing from an eye, hearing from an ear, etc. The better its purpose is realized, the more organized is the system. We need criteria to characterize biological order; we need to explain the existence of selection pressures which account for the advantages gained when our purpose is realized with great efficiency. The index for evolution may be entropy production and the level of interactions with the outside world. There are probabilities we can assign to the occurrence of no event and, on the other hand, to one event. In other words we can use the number one or zero, to represent the fact that the event has, or has not, happened. This number, one or zero conveys exactly one bit of information (bit: a contraction of the words binary and digit). A binary number provides an amount of information which is equal to the entropy of the process generating the event. The information transmitted links to the change in the state of organization of the system. For example, during the first stage of oil exploration, an oil company will determine that one of its new wells will or will not produce oil. In the second stage, the directors of the company will or will not vote a dividend. We can define probabilities in terms of (1) success in finding oil and a dividend is paid; (2) success and no dividend is paid; (3) no success in finding oil and a dividend is paid and (4) no success and no dividend is paid. We can measure the information (in entropy units) available in terms of these probabilities and ask how much information has been gained through the knowledge that the well was successful, calculated from Shannon's formula.

Decentralization is a process by which decision-making authority is delegated down the organizational hierarchy. The effectiveness of such decentralization is adduced from the levels at which decision-making authority is placed and the relative importance of the decisions. We can locate the center of decision-making authority within an organization; but we are also concerned with the dispersion of the decision-makers, this being part of the index of decentralization. Entropy is a useful measure of dispersion and decentralization, where entropy of an organization is defined as the probability of a decision being made multiplied by the logarithm of that probability. Summing these numbers over all levels of the organization is what is called Shannon's entropy formula, the uncertainty index in decision-making. If all decisions were made at one level of an organization there would be no uncertainty as to where decisions originate: an example of complete concentration of decision-making authority. Alternatively, the greatest uncertainly would occur when it was equally likely that a decision could be made at any level of organization (a maximum dispersion of decision-making). Entropy calculated from Shannon's formula is increased when the uncertainty of a situation also increases (more dispersion of decision-making authority), attaining a maximum when all probabilities are equal. Entropy and dispersion may increase as the number of levels of the organization is raised. The difficulty with using Shannon's formula of course resides in the stygian problem of finding the probability that a decision is rendered at a given level. If everyone is independent, we can relate probabilities to frequencies — how often decisions are made at each level and then use entropy as a measure of dispersion in the analysis of business and economic data. Hildenbrand and Paschen were early believers in entropy in the analysis of economic data, using it as a measure of industrial concentration, though Thiel applied it more extensively to industry sales, foreign trade and corporation balances sheets. Thus should the calculations yield an entropy value for imports of leading commodities of 2.76 for 1950 and 2.86 for 1960, we may conclude that disorder in the distribution of imports increased during this decade or that there was greater equality (likelihood) of importing food, petroleum, paper, etc. In the investment field we could assign a probability that various securities might yield given returns over a period of time and compute by Shannon's formula the entropy distribution of stock prices. Or how is the world trade distributed among countries. How are assets distributed on a balance sheet? The results of these calculations have been interpreted syncretically as signifying a freedom of choice, uncertainty, disorder, information content and information gains or losses. A popular use of entropy is in the analysis of market structure. In a monopoly situation the buyer has no choice as to the firm to be patronized, the lack of choice is reflected in the entropy measure (a value of zero for a perfect monopoly). In a two-firm duopoly situation — equal sized companies — the random buyer is equally likely to patronize either firm and has only one decision: to buy from one or the other. Shannon's formula would in this case yield a number equal to 1.0. Where the businesses are of unequal size, the more unequal they are, the smaller the entropy value. In these calculations, entropy equates with the average number of dichotomous decisions the buyer must make, where the goal is to minimize the number of decisions. It becomes the degree of uncertainty as to which firm is to be chosen by the random customer. As such, entropy reflects the degree of competition. Entropy has been used in accounting, to quantify the putative loss of information when items in a financial statement are combined. The entropy difference before and after combining a pair of entries tells us something about the information loss in aggregation. Information here means the elimination of uncertainty as to which of several messages has been transmitted. Or the freedom of choice we have in selecting a particular message from a set of all possible messages. Entropy is a measure of information in the sense that it focuses on the uncertainty surrounding a transmitted message. The problem of aggregation in accounting and in financial analysis is primarily that of providing the user with sufficient insight into the financial condition of the firm or the market. Rearranging and combining items in various ways may reduce the user's insight and cause a different meaning to be attached to the message. We measure disorder, dispersion or centralization, in entropy terms, using Shannon's formula.

Davis and Lisiman suggested money as the analog of economic entropy, a concept transcending the Leontief static input-output equations (economic processes completely described by the flow of commodities into and out of the system). Georgescu-Roegen recommended some biological homologies to account for changes with time that Leontief ignored. The economic process is entropic; it neither creates nor consumes matter or energy but only transforms low-entropy systems to higher levels. Money constitutes the economic equivalent of low entropy. We seek a conversion factor to find the entropy equivalent of economic value. Entropy changes in closed chemical systems have been traced for about one hundred and twenty years; for open systems it is ninety years. Extension of the open system analysis into situations far from equilibrium earned Prigogine the 1977 Nobel Prize in chemistry. Even before this, entropy and biological processes were linked; later entropy and business activity were joined. Perhaps the earliest work in irreversible thermodynamics was Thompson's examination of steady state open systems: the creation and maintenance of a concentration difference across a film of liquid with a temperature difference. The heat flow caused the concentration gradient; the concentration gradient implied order (a higher order than before and less entropy). All caused by the heat flow. Onsager won a Nobel Prize for his thermodynamic work on irreversible processes not far from equilibrium, driven by modest forces. Prigogine and his coworkers went beyond Onsager and developed theories for the appearance of new structures far from equilibrium, beginning with the decomposition of the entropy production term into two parts: from the internal irreversible reactions and the interactions of system and environment. Social organizations as diverse as informal friendships and cultures have definable conditions of order and efficiency just as surely as do steam engines. Three basic flow categories have been identified which cross corporate boundaries: matter, energy and information. Specific flows into the system include, among others, raw materials, supplies, equipment, services, work by employees, information and capital, usually expressed in monetary not entropy terms. The flows out of a manufacturing operation are products, by-products, wastes, information and dividends. There are also pressures from the environment: government regulations and actions of competitors. Internal processes include the aging of employees and equipment, generation of novel information, changes in organizational structure and conflicts among employees. The search for a relationship between value and entropy began around 1880 when money became socio-biological energy, a simplistic approach to an esoteric question. Suppose we are to boil water using alternatively, wood, natural gas and coal as the fuel. The entropy change of the water, in going from liquid water to steam is the same if we begin at the same temperature and pressure and end similarly for each of the three different burning processes. But the cost of the product, steam, depends on which fuel is used and hence the entropy-money conversion cost is not fixed. Ostwald proposed that economic value depends only on the energy stored and available in the wood, natural gas or coal but it should be pointed out that the efficiency of the burning process is also important. (The value of the steam may depend on whether the wood is burned in an open fire or in a controlled draft furnace.) Entropy changes can be calculated for: water to ice transformations; carbon dioxide gas to solid dry ice changes; the graphite-diamond correspondence and helium gas-helium liquid phase inversions. Depending on the energy, equipment and labor cost we can produce empirical correlations to describe an entropy-money conversion factor. If these factors were readily available, we could make an entropy balance around a corporate open system and establish its entropy exchange with the environment. Since the late 1940's entropy calculations using Shannon's formula have analyzed the dispersion of decision-making in the business world, between groups and within groups in a hierarchical organization.

Hahn attempted to establish a relationship between the usual corporate performance indicators such as profits, sales, etc., and informational entropy. The goal was to correlate fiscal performance with internal hierarchal structure (the table of organization). Corporate structure was written as an assembly of blocks, such as those for the President, and the Vice President for Sales. Each block was characterized by a power factor containing two parameters: (1) the fraction of the total corporate budget controlled and (2) the hierarchial level below the President (where the top rung is 1, Vice Presidents are 2, etc.). Thus the measure of power each organizational block carries is directly related to its budgetary responsibility and inversely proportional to the distance from the ultimate decision-maker, the President. Changes in fiscal performance were mapped along with entropies calculated by Shannon's formula. This preliminary study of Hahn's suggests that an optimum structural configuration may exist, where performance and productivity can be maximized within an organization by establishing which internal structure is most conductive for optimum information flow. Shannon's formula is a useful tool here for deriving entropy data because: (1) maximum entropy seems to correlate well with the most highly disordered state; (2) increasing the number of hierarchical blocks raises the organizational entropy; (3) we can glean from entropy a measure of concentration and dispersion. An inefficient and disorganized company should evidence high entropy and high operating expenses. For the jejune company there will probably be a jump in profits as size and structure increase — up to the point where size and structure begin to work against an efficient operation. There can be, with some structures, too little informational entropy (the capacity to transmit information) where insufficient capacity to transmit information means too much information is being stored and not used. This excess of stored information (calculated from Shannon's formula) could likely garble the lines of communication between the various departments of a corporation, manifested by tedious administrative and clerical detail as well as excessive interdepartmental formality and reserve. The result could be a hierarchy so complex and rigid that the system's ability to disseminate information quickly to its components and to respond with flexibility in the face of change is seriously undermined.

Living systems are born, mature, senesce and die, all performing their little dance of life within a finite time frame. It is easy to plot their course for we know there is a beginning, middle and an end. For so-called inanimate systems such as corporations, countries and civilizations, it is not as simple a matter to identify the opening event, birth, or the denouement, death, with such clarity. What of mergers? How to account for human interference, which alters the course of history? Who can say that a country has died when it is conquered? Do civilizations really expire?

A nascent civilization has surmounted the first and highest hurdle but will it then go from strength to strength? And for how long? The growth of a society can be measured in terms of its increasing power of self-determination; the future lies in the hands of the creative minority. A civilization that attempts some extraordinary tour of force may find itself not defeated outright, but arrested in a state of immobility, its energies absorbed in meeting the single great challenge. The Central Asian Nomads were condemned to this fate: they successfully mastered the problem of adaptation to the harsh exigencies of life on the steppes, but in doing so they became the slaves of their environment, unable to make any fresh creative advances. The movement of challenge-and-response becomes a self-sustaining series if each successful response provokes disequilibrium, requiring new creative adjustment. Society is a network of relations, the interaction of two or more agents. A society is the medium of communication through which humans interact. The process by which growth of civilizations is sustained is inherently risky: the creative leadership of a society has to carry along the uncreative mass. The ultimate failure of creativity seems to stem from its successes, whereby we seem to become lazy or self-satisfied or conceited. These numinous successes inspire others to follow but often create a requirement for dull obedience, such as soldiers in an army. In some civilizations, an army on becoming demoralized degenerates into rebellion.

Spengler argued that a civilization is comparable to an organism, subjected to the same processes of childhood, youth, maturity and old age as a human being or any other living thing. He claimed every civilization, every archaic age, every rise and downfall has a definite time frame which is always the same and which always recurs. The breakdown is connected with the periodic rhythm of life — in animals, the vegetable kingdom and the so-called inanimate world. Do you accept the premise that civilizations have met their death not from the assault of external and uncontrollable forces but by their own hands? We become complacent; we lose the Promethean élan of the unstable equilibrium in which custom and mimesis are never allowed to become entrenched. The forces acting upon the arrested society demand so much of their energies simply to maintain the position already attained; there is nothing left for reconnoitering the course ahead. The nemesis of creativity in a civilization is the idolization of an ephemeral self or institution. Breakdowns of civilizations may not be inevitable, but in the process of disintegration, if allowed to fester, they show some similarities with the myriad of social systems. The masses became estranged from their leaders, who then try to cling to their positions by using force as a substitute for their evanescent power of attraction. A dying society molders in a climate of violence, seen by the victors as a cataclysm in which the forces of evil are destroyed and a new age of peace inaugurated. Demoralized by their failures, the people of a disintegrating society resort to parodies of the creative inspiration they seem to have lost. The cults which arise are generally barren and in general represent an attempted escape from an intolerable world. In passing from the breakdown of a civilization to its disintegration, we should not readily assume that this sequence is automatic and unalterable — that once a civilization has broken down, it must inevitably drive towards disintegration and dissolution. In the disintegration of civilizations, the perpetual variety and versatility which are hallmarks of growth yield a noisome uniformity and un-inventiveness. After the failure to meet the challenges, the old unanswered challenges reappear wraithlike to present themselves more insistently and in even more virulent form until at last they dominate and obsess and overwhelm the society. Thus, the disintegration of a civilization, like its growth, is a cumulative and continuous process. In essence, the loss of harmony between previously coexisting elements in a society leads ineluctably to social discord. There are vertical schisms among geographically segregated communities and horizontal schisms operating on geographically intermingled but successfully segregated classes. The horizontal schism of a society along lines of class appears at the moment of breakdown, a distinctive benchmark, absent during the growth phase. The disintegration begins when status is questioned, and hostile fragments arise. The ruthless pursuit of incompatible class interests shatters the social pyramid and creates new structures of oppression. In a growing civilization the creative minority can exercise its powers of attraction upon neighbors beyond its borders as well as upon the internal community. The diddling of a disintegrating society emboldens its neighbors, who now become a menace; a weakened Roman Empire from the fourth to the sixth century was under continual attack on its northern front by successive waves of Huns, Avars, Teutons and Slavs, pilling up against each other and eventually overrunning the whole of Rome's Empire in the West.

When a society is disintegrating, the dominant minority tries to preserve its threatened power by uniting the warring nations (the external proletariat) into a universal state or empire, a seemingly single civilization. The creation of a universal state checks the headlong decline of a disintegrating civilization; such states are endorsed eagerly by those who believe their empire is destined to be as immortal as the gods who have ordained it. This conviction is sustained by the vainglorious assertion that their universal state will embrace the whole world, leaving of course no external force to threaten it. Universal states unify one civilization, but also embrace portions of alien societies. For the rulers of universal states a network of communication is an indispensable instrument of military and political control. In a universal state the capital city derives enormous prestige from its status as the seat of government. The degree of efficiency attained by imperial administration varied considerably among universal states. Universal states are essentially negative institutions; they arise after, not before, the breakdown of the civilization to which they purport to bring political unity. They are the products of dominant minorities: the once creative minorities that lost their Promethean power. Universal states seem to be possessed by an almost demonic craving for life; their citizens believe passionately in the immortality of their country or institution. However long the life of a universal state may be drawn out, it has always proved to be the last phase of a society before its extinction. Its goal is the achievement of immortality, a vain effort to thwart the economy of Nature.

Civilizations are brought to grief by their own faults and failures, not always by external agencies. A society in the process of dissolution is usually overrun and finally liquidated by so-called barbarians from beyond its frontiers. The barbarians are the brooms which sweep the historical stage clear of the debris of a dead civilization, much as the phagocytes of our bodies perform their task for us. The weakened civilization, temporarily reconstituted in a universal state surrounds itself with rigid political frontiers which repel the attractive forces of the barbarians. Not able to imitate the culture and art of the universal state, frustrated, the barbarians attack. Hadrian's Wall was visible evidence of the Roman Empire's attempt to insulate itself from the barbarian tribes of northern Britain and to protect against invasion. China's Great Wall, is tangible evidence of an embattled civilization's attempt to establish rigid lines of defense against outer barbarians.

CHAPTER 12

Informational Entropy and Shannon

The concept of entropy has been widely discussed in many scientific and social arenas and its application to organizational structure is a logical extension of the work done previously by Prigogine, Georgescu-Roegen, Shannon, Quastler, Horowitz and others.

One of the most famous uses of the entropic concepts in chemistry was by Ilya Prigogine, recipient of the 1977 Nobel Prize in chemistry. He correlated non-equilibrium phenomena and disorder (entropy). He studied systems near equilibrium and systems with minimum entropy production. However, beyond a certain critical distance from equilibrium, entirely new structures could emerge. These new systems, far from equilibrium, Prigogine referred to as dissipative structures. In addition to his work in chemical thermodynamics, Prigogine has also been an innovator in the field of social thermodynamics. He has adapted many of the principles of thermodynamics to social organizations, recognizing that each organization is itself an open system. Another contributer to the interdisciplinary analysis of open systems was economist and thermodynamicist, Nicolas Georgescu-Roegen, who believed that the notion of entropy had great utility in economics, that the earth is an open system with irreversible processes occurring within it.

The mathematical definition of informational entropy was derived by C. E. Shannon in 1949. With molecules in motion, colliding and rebounding, different molecules will occupy a given space at various times and hence many molecular arrangements, called microstates, are possible. We associate the concepts of greater freedom, uncertainty, and more configurational variety with an increased number of microstates and higher entropy. If one had to guess where a particular molecule would be at a given time, the probability of error would be greater in the higher entropy state.

Thus ordering of a system implies lower entropy, which carries with it a certain reliability and smaller probability of error.

From this we know that as the number of microstates, W, increases, the state entropy, H, increases. Hence we can write,

$$H = K \log_a W, \tag{1}$$

where a is the logarithmic base.

If we assume that all the microstates are equiprobable, then the probability of achieving each individual microstate, p_i, is simply one out of the total number of the microstates, W, or

$$p_i = 1/W \tag{2}$$

or

$$H = -K \log_a (p_i). \tag{3}$$

We can extend this idea to non-equiprobable systems with the use of the Expectation Value, E_x, which is by definition the probability of each outcome, p_i, multiplied by the value of that individual outcome, X_i, summed over all possible outcomes, as indicated in Eq. (4).

$$E_x = \sum_i p_i X_i. \tag{4}$$

With the probability, p_i, and $X_i = H_i = -K \log_a (p_i)$, from Eq. (3), E_x (also called S, the entropy of the system) is

$$S = -K \sum_i p_i \log_a p_i. \tag{5}$$

Thus, Eq. (5) expresses the entropy of a system in terms of probabilities. It takes the concept of entropy from the thermodynamic setting to the domain of general probability theory. It can be shown that S in Eq. (5) achieves a maximum value if and only if all the p_i are equal. If for convenience, the constant K is taken to be unity, then Eq. (5) reduces to

$$S = -\sum_i p_i \log_a p_i. \tag{6}$$

This is Shannon's formula for informational entropy.

The Meaning of Stored Information

Weaver states that the word information in communication theory relates not so much to what you say, as to what you can say. Potential message variety, freedom of choice, and a large vocabulary are the desired ends of communication and information transmission. A library obviously contains stored information. The information is stored in a linear sequence of symbols organized to the constraints of a language. The subsequences are organized into books and periodicals, and these are carefully ordered on shelves. Everywhere order and constraints are associated with the information storage process. This is a state of low entropy. If we take each page of each book, cut it into single-letter pieces and mix them in one jumbled heap, the entropy would increase and stored information would decrease and potential information has increased. In entropy terms, stored information is the divergence from the state of maximum disorder (when all p_i terms are equal). In other words, stored information is the difference between entropy content for the equally probable state, S_{max} (maximum disorder) and that for the unequally probable present, S, and is denoted by D. Therefore, stored information is

$$D = S_{max} - S. \tag{7}$$

Entropy and Corporate Structure

Shannon's informational entropy formula has in the past found application by Gatlin, who computed genetic stored information, by Horowitz and Horowitz, and Herniter in marketing, by Lev in accounting, by Thiel and Georgescu-Roegen in economics, by Philipatos and Wilson in securities analysis, and by Murphy and Hasenjaeger in organizational decentralization.

These authors extended the definition of p_i from the probability that a system will be in a particular microstate to such related considerations as: (1) the probability that a customer will purchase a product; (2) the degree of competition in the marketplace; (3) a measure of the dispersion in a securities portfolio; (4) the degree of market share; (5) the degree of organizational decentralization; and (6) the bits of stored genetic information.

One can generate a Power Index, P_i, to be used in Shannon's formula, analogous to p_i, in order to characterize the overall structure of a corporation. Since each unit controls those beneath it, one can also define a

Cumulative Power Index for each unit, C_i, where

$$C_i = P_i + \text{sum of all } P_i \text{ controlled by this unit, } i$$

Finally, a fractional Cumulative Power Index, f_i, is introduced as the unit's C_i divided by the sum of all C_i in the organization. Thus Eq. (6) can be transformed to

$$S = -\sum_i f_i \log_2 f_i. \tag{8}$$

We can define an entropic distance from disaster, D, for the actual structure, by Eq. (9)

$$D = S_{\text{max}} - S, \tag{9}$$

and the distance from disaster for an ideal structure, D^0, as

$$D^0 = S_{\text{max}} - S^0, \tag{10}$$

where S^0 is the entropy of the ideal structure, and a structural efficiency, η, by Eq. (11)

$$\eta = [(S_{\text{max}} - S)/(S_{\text{max}} - S^0)] \times 100. \tag{11}$$

The distance from disaster, D, is also stored information, Eq. (7), as well as Excess Entropy for the inanimate system, the corporate-style organization,

$$EE = S - S_{\text{max}}, \tag{12}$$

where S_{max} = maximum entropy content for the organization, if all the units have the same budget, are at the same level, and are completely independent: a prescription for disaster, ($S_{\text{max}} = \log_2 n$ and $f^{\text{disaster}} = 1/n$, where n is the number of units in the organization.)

From Eqs. (7), (8) and (12) and recognizing that Excess Entropy Production (EEP) is expressed as the product of thermodynamic forces and flows, we can write

$$EEP = \delta S \delta EE, \tag{13}$$

where

J, thermodynamic flow $= S$
X, thermodynamic force $= EE$

and where previously, for the living system, flow $= r$ (chemical kinetics) and force $= A$ (chemical affinity). It can be shown that

$$EE = S - S_{max} = S - \log_2 n \tag{14}$$

and

$$EEP = \left[\sum_{j}^{n} (f_i - 1/n)(1 + \log_2 f_i) \right]^2. \tag{15}$$

For each year of a corporation's history, f_i can be computed for each unit and the summing process of Eq. (8) accomplished to produce S and then efficiency, η, from Eq. (11). The EE is obtained from Eq. (14) and EEP from Eq. (15). Thus EE and EEP longevity tracks can be constructed, just as they can be (and have been) for the living system.

More Details

Hershey used a modified version of Shannon's informational entropy formula in the analysis of corporate-style organizations. This is given by Eq. (16)

$$S = - \sum_{i=1}^{n} f_i \log_2 f_i, \tag{16}$$

where f_i = fractional cumulative power index, obtained from P_i, where P_i = power index = the budget of each unit divided by an integer $(2, 3, 4 \ldots)$ which represents the level below the president (level 1).

C_i = cumulative power index = P_i + (the sum of all P_i controlled by this unit) and

$$f_i = C_i \Big/ \sum_{i=1}^{n} C_i.$$

We can define Excess Entropy (EE) for the inanimate system, the corporate-style organization

$$EE = S - S_{max}, \tag{17}$$

where S_{max} = maximum entropy content for the organization, if all the units have the same budget, are at the same level, and are completely independent. A prescription for disaster.

It was shown previously by Patel that

$$S_{max} = \log_2 n \tag{18}$$

and

$$f^{disaster} = 1/n, \tag{19}$$

for the organizational configuration leading to S_{max}.

Recognizing that EEP is expressed as the product of thermodynamic forces and flows, we can again write

$$EEP = \delta S \delta EE, \tag{20}$$

where, as before, J, thermodynamic flow $= S$; X, thermodynamic force $= EE$.

From Eqs. (17) and (18) we obtain

$$EE = S - S_{max} = S - \log_2 n \tag{21}$$

and

$$EEP = \delta S \delta (S - \log_2 n) = (\delta S)^2, \tag{22}$$

where

$$\delta \log_2 n = 0 \quad \text{since } \log_2 n \text{ is a constant.}$$

From Eq. (16)

$$\delta S = \delta \left[\sum_{i=1}^{n} (f_i \log_2 f_i) \right]$$

$$= \sum_{i=1}^{n} (f_i \delta f_i / f_i + \log_2 f_i \delta f_i)$$

$$= \sum_{i=1}^{n} (\delta f_i)(1 + \log_2 f_i). \tag{23}$$

From Eq. (19) we obtain

$$\delta f_i = f_i - f_i^0 = f_i - f^{disaster} = f_i - 1/n \tag{24}$$

and

$$EEP = \left[\sum_{i=1}^{n} (f_i - 1/n)(1 + \log_2 f_i) \right]^2. \tag{25}$$

CHAPTER 13

The Corporate Structure

Some say the organization of the future will be a confederation of internal entrepreneurial teams. To be avoided are bureaucratic organizations where work life more closely resembles life in a totalitarian state than in a free nation. In today's complex and intelligence intensive world economy, it is becoming obvious that, in corporations as in nations, totalitarian governance and overbearing bureaucratic management may be incompatible with high performance. The greater the freedom of its parts, the faster an organization can learn. The system acquires a permanent speed and adaptability.

Max Weber, who launched the systematic study of bureaucracy in the late 19th century, saw the bureaucracy as the most efficient possible system. Bureaucracy gained prominence because it worked for many of the needs of the industrial age. With its pyramid form, absolute power at the top and divided tasks, the bureaucracy gives responsibility to sub-bosses on the lower levels. (Some organizations had as many as twelve layers of management between the CEO and the workers.)

With the recognition of the limitations and hindrances of the pyramidal hierarchical corporate structure, others are supporting more relaxed relationships among their internal and external units. This provides a more "organic" feel to their operations. They evolve, and the boss is simply part of the system. The architecture of intelligent organizations will be flexible, shifting appropriately to meet new challenges and responding to local situations. The challenge then is, upon demand, to depart from the classical pyramid, to reach more manageable sizes. To reach dimensions consistent with a human's ability to understand the system.

Many deficiencies of the bureaucracy are compensated for by generating a "shadow" or informal organizational structure, whereby it is possible to make connections or interactions across distant boundaries of the bureaucracy. This bypassing of your boss philosophy may work in the short run, but it will clearly cause confusion in information flow. In aged bureaucracies, this informal organization is groaning and overloaded. The challenge is to balance the good and bad of the shadow structure, against the best and worst of the hierarchical organization. Whatever the final choice, it is known that for an organization to be intelligent, it must be aware of its surroundings, and aware of itself.

Bureaucracies are often unable to sense what is going on around them. Frontline employees are focused on the rules and the chain of command. As information travels upward, each level sugarcoats or hides what it thinks might be sensitive or embarrassing. That information which reaches the top often does not reflect reality. The problems with information flow in bureaucracies reside not just within the upward flow, but also within its lateral information flow, which is restricted. With entrepreneurial sub-units however, a bureaucracy can learn faster both because it is more highly connected and because pluralism and choice creates a hunger for new ideas and a more honest dialogue.

In social circumstances, a circle is the natural geometrical form for communications. Whoever heard of a family square, or a rectangle of friends? What is important, whatever the geometrical shape of the organization, is to be open to a variety of communications options, to allow an evolution or differentiation from here to there. This allows more degrees of freedom, and more ways to handle problems.

Complexity and Communications

The caution is to avoid the separation of authority and responsibility, while allowing the process of disintegration and reintegration to proceed continuously. Otherwise, the response to rapid change will be inadequate: we will provide solutions for problems which no longer exist. This surely would lead to chaos, where the parameters against which we are working have been changed. For example from the mathematics of chaos, the equation $Y = 4\lambda(1 - x)$ will yield stability, complexity, or chaos, depending on the value of the λ parameter.

Various Geometric Shapes Available for a Table of Organization

Characteristics of Four Different Networks

	Chain	Wheel of Star	Circle	All Channel
Example	Chain of Command	Formal work group	Committee or task force: autonomous work group	Grapevine: Informal Communication
Centralization of power and authority	High	Moderately high	Low	Very Low
Speed of communication	Moderate	Simple Tasks: Fast Complex Tasks: Low	Members together: Fast Members isolated: Slow	Fast
Accuracy of communication	Written: High Verbal: Low	Simple Tasks: High Complex Tasks: Low	Members together: High Members isolated: Low	Moderate
Level of group satisfaction	Low	Low	High	High
Speed of decisions	Fast	Moderate	Slow	—
Group commitment to decisions	Low	Moderate	High	—

In a hierarchy, communications problems increase with size. Problems also arise from interactions which bypass the usual chain of command. Vertical structures contaminate the quality of information flow more readily than do horizontal configuration. Religious establishment seems guided by strict hierarchies, where various units within the organization merge at power nodes. We also find power nodes in living systems, in the synapse where separate neural cells are joined, and information facilitators (neurotransmitters) operate, and information flow exists through permeable membranes. The structure in a hierarchy, living or otherwise, may be uninodal or multi-nodal, depending on its size and complexity.

Tables of Organization

Institutional Level (upper part): top level managers and administrators (long range planning, policy making, and managing the organizational environment).

Managerial Level (center part): middle management (coordinate the internal activities of the organization).

Technical Core (bottom part): operating employees (deal with short run objectives).

Of the numerous attempts to quantify the effectiveness of group structures, many have examined the degree of structural centrality. Max Weber associated the hierarchy with structural centrality and rationality. He attributed these characteristics to the hierarchical structure:

- Specialized and division of labor
- Rules of procedure
- Authority
- Impersonality of office
- Employment and promotion by merit

Some believe that decentralization restricts innovation and growth to existing projects. Some believe that tall (vertical) structures require close control of subordinates and more personal contact between superiors and subordinates. Some believe that tall (vertical) growth requires a strict chain of command and so-called unit commanders. Some believe that flat (horizontal) organizations develop specialized units.

Beyond these paradigms, we have seen evolve, the matrix organization. Tasks require specialists with possibly two or more supervisors.

In a matrix organization, the product is controlled by a manager for that product. Employees work on different products as part of different teams. There are project teams and product groups, and employees develop specialized knowledge. A matrix organizational structure can respond quickly to market fluctuations. There is the ability to balance on product against another.

The corporate planners need to weigh the following broad choices for their bureaucratic model:

- The Rational Bureaucracy (most common)
- Collegial Consensus (small groups, team work)
- Pluralistic (a political compromise, a mixture of the above)

Some who study organizational structure have concluded that size affects organizational activity, that there is a relationship between group size and complexity. They defined interactions as relationships between pairs of units, and suggested that these interactions also increase complexity.

The concern for efficient information flow and the disorder of choices, led to Shannon's equation for characterizing this information flow,

$$H = \sum p_i \ln p_i$$

where

H = A measure of the concentration of authority. Small H values mean high concentration of authority (low uncertainty and low disorder)

p_i = The power of each unit within the structure

Saying power is concentrated means the p_i are not the same for all units. Maximum disorder implies each unit gets exactly the same choices (same p_i). The condition of minimum disorder is when one unit contains all the authority or power.

Today we consider the corporate organization as an open system in the thermodynamic sense, with information flow possible across unit boundaries. Energy and material also diffuse into and out of the units. And there can be redundancies in the structure, in its parts, functions and change processes.

Informational Entropy, Shannon's Approach

The mathematical definition of informational entropy was derived by C. E. Shannon in 1949. The relationship between informational entropy and thermodynamic entropy can be seen by a classic thermodynamics example.

Suppose we have two distinguishable gases A and B, separated by a removable partition in a thermally insulated container. This is an ordered state since each gas is confined to only half of the box.

When the partition is removed, the gas molecules will diffuse and intermix until a homogeneous state is reached. The entropy content of system has increased. We know that these molecules will not spontaneously return to the original partition configuration.

Divide the container into imaginary compartments or microstates and suppose we can tell which molecules are in a particular compartment or microstate at a given time. The exact number of microstates possible defines a thermodynamics probability and is denoted by letter, W. After removal of the partition, a greater number of microstates is possible than before, since all molecules are free to roam over the entire box. Without the partition, there is great uncertainty as to where a particular molecule will be and the thermodynamic entropy has increased. Thus with the state of higher entropy we associate the concepts of greater freedom, uncertainty, and more configuration variety and an increased number of microstates.

More on Entropy, Structure and Shannon

The elements of uncertainty and configuration variety are important in bridging entropy and information concepts. Uncertainty implies a higher probability of error.

Higher Entropy	Lower Entropy
Random	Nonrandom
Disorganized	Organized
Disordered	Ordered
Configurationally variety	Restricted arrangements
Freedom of choice	Constraint
Uncertainty	Reliability
Higher error probability	Fidelity
Potential information	Stored information

$$H = K \cdot W, \tag{1}$$

where H denotes entropy, K is an arbitrary constant and W is the thermodynamic probability (or the number of microstates).

If the entropy of system A plus that of system B is equal to the entropy of the composite system, AB, then

$$H_A + H_B = H_{AB} \tag{2}$$

and from Eq. (1)

$$W_A + W_B = W_{AB}. \tag{3}$$

But from the theory of probability, the number of microstates of a composite system AB is the product of the microstates of A and B, not the sum ($W_{AB} = W_A W_B$).

Boltzmann showed that when numbers expressed as powers of the same base are multiplied, their exponents are added.

$$a^p \cdot a^n \cdot a^m = a^{p+n+m}, \tag{4}$$

and that logarithms have similar properties,

$$\log pnm = \log p + \log n + \log m. \tag{5}$$

Therefore, he used the ideas expressed in the Eqs. (1) through (5) by writing.

$$H = K \log_a W, \tag{6}$$

where a is the logarithmic base.

If we assume that all the microstates are equiprobable, then the probability of achieving each individual microstate, p_i is simply one out of the total number of microstates, W, or

$$p_i = 1/W \quad \text{and} \quad W = 1/p_i. \tag{7}$$

Substituting for W in Eq. (6), we get

$$H = K \log_a (1/p_i) \tag{8}$$

or

$$H = -K \log_a (p_i). \tag{9}$$

As shown previously, we can extend this idea to non-equiprobable systems with the use of the Expectation Value, E_x, which is by definition, the probability of each outcome, p_i, multiplied by the value of that individual outcome, X_i, summed over all possible outcomes, as indicated in Eq. (10)

$$E_x = \sum_i p_i \cdot X_i, \tag{10}$$

which yields Eqs. (11) and (12)

$$S = -\sum_i p_i \left(-K \log_a p_i\right) \tag{11}$$

or

$$S = -K \sum_i p_i \log_a p_i = -\sum_i p_i \log_a p_i, \tag{12}$$

if we neglect K.

Units of Informational Entropy

Though the units of information entropy can be arbitrary, it is convenient to use logarithms to the base 2. Hence when there are only two choices, p_i is 1/2, $\log_2 2$, is unity, and S in Eq. (12) is also equal to unity. The unit of information (or S) is called a "bit", being a condensation of "binary digit", which was suggested by Tukey. Thus "bit" represents a two choice situation which has unit information, i.e., from Eq. (12).

$$\begin{aligned}
S &= -\sum_1^2 (1/2) \log_2 (1/2) \\
&= -[(1/2) \cdot (-1) + (1/2) \cdot (-1)]] \\
&= 1 \text{ bit}
\end{aligned}$$

Finally, Eq. (12) can be conveniently written as

$$S = -\sum_i^n p_i \log_2 p_i. \tag{13}$$

The Meaning of Stored Information

Weaver states that the word information in communication theory relates not so much to what you say, as to what you can say. A library obviously contains stored information. This is a state of low entropy. In entropy terms stored information is the divergence from the state of maximum disorder. Stored information is the difference between maximum entropy content, S_{max} (maximum disorder), and that for the entropy of the present, and is denoted by D. Therefore, stored information is

$$D = S_{max} - S$$

or

$$\frac{D}{S_{max}} = 1 - \frac{S}{S_{max}} \quad \text{or} \quad \frac{S}{S_{max}} = 1 - \frac{D}{S_{max}}. \tag{14}$$

which approximates stored information.

Application of Informational Entropy in Business

The concept of informational entropy is becoming popular in marketing, subject to two fundamental presumptions regarding probability. They are (1) a firm's market share represents the probability, p_i, that a customer will purchase a product from a particular firm and (2) the customer's product choices are unprejudiced. For instance, $p_i = 1.0$, signifies no freedom of choice, a monopolistic situation (no competition) which would yield from Eq. (13) zero informational entropy. In marketing, informational entropy serves to measure the degree of uncertainty in the market, i.e., it reflects the degree of competition among the firms.

In addition to its applicability in marketing, Lev has also found a place for informational entropy in accounting. He suggested that information entropy could be used to measure the loss of information that occurs when items on a financial statement are combined. Lev also believes that his approach can be of assistance to people who are doing aggregation on income statements.

Theil has employed Shannon's formula in a wide variety of situations in economics. A number of business statisticians have stressed entropy in their analysis of securities portfolios. G. C. Philippatos and C. J. Wilson felt that informational entropy values derived from Shannon's formula were ideally suited for measuring the dispersion in security portfolios because the higher the entropy, the greater was the dispersion.

I. Horowitz, proposed that entropy is a meaningful index of industrial competition. A higher entropy value was synonymous with greater competition. A. R. Horowitz examined the American brewing industry, observing that as a firm's fractional market share, p_i, approached unity, entropy tended towards zero (a monopolistic position).

Extensive application of Shannon's formula was undertaken by Murphy and Hasenjaeger. They considered informational entropy a measure of decentralization, as a process by which decision-making authority is delegated down an organizational hierarchy.

Calculations of entropy changes in open systems were reported in 1896, entropy production in 1911. Clausius coined the word, entropy, from the Greek language to mean transformation and defined it as the increment of energy added to a body as heat divided by its temperature, during a reversible process. Brillouin showed a connection between entropy and information. Dowds developed his own method of applying entropy concepts to the examination of the data from oil fields, as did others in land use planning and in prediction of travel between different communities and in Nigeria to analyze water runoff. It was the work of Shannon that more firmly connected information and entropy.

Shannon's formula connects informational entropy to the probability that the system is in a certain structural configuration. A number from one to a hundred can always be identified in seven guesses (which can be answered by "yes" or "no"). To do this one needs to make guesses in such a way as to eliminate one-half of the remaining range. By applying Shannon's formula we can calculate the informational entropy for this process; we get the number of guesses required to find the unknown number, or the amount of entropy information associated with the process. The relationship between entropy and information provides further clarification of the fundamental principle of the living process: increases in entropy can be viewed as a destruction of information. A living, open organism must be able to maintain its organized state against the forces of disorganization; it must continue homeostasis and a purposeful behavior.

Describing self-organizing processes requires us to define and quantify the notion of order; we need a model to account for the transition from one slot in the hierarchy to the next. The index for evolution may be entropy production. There are probabilities we can assign to the occurrence of no event and, on the other hand, to one event. In other words, we can use the number one or zero to represent the fact that the event has, or has not, happened. This number, one or zero conveys exactly one bit of information. A binary number provides an amount of information which is equal to the entropy of the process generating the event. The information transmitted links to the change in the state of organization of the system.

For example, during the first stage of oil exploration, an oil company will determine that one of its new well will or will not produce oil. In the second stage, the director of the company will or will not vote a dividend. We can define probabilities in terms of (1) success in finding oil and a dividend is paid; (2) success and no dividend are paid; (3) no success in finding oil and a dividend is paid and (4) no success and no dividend are paid. We can measure the information (in entropy units) available in terms of these probabilities and ask how much information has been gained through the knowledge that the well was successful, calculated from Shannon's formula.

Decentralization is a process by which decision-making authority is delegated down the organizational hierarchy. The effectiveness of such decentralization is adduced from the levels at which decision-making authorities are placed and the relative importance of the decision. We can locate the center of decision-making authority within an organization; but we are also concerned with the dispersion of the decision-makers, this being part of the index of decentralization. Entropy is a useful measure of dispersion and decentralization, where entropy of an organization is defined as the probability of a decision being made multiplied by the logarithm of that probability. Summing these numbers over all levels of the organization is what is called Shannon's entropy formula. If all decision were made at one level of an organization there would be no uncertainty as to where the decision originates: an example of complete concentration of decision-making authority. Alternatively, the greatest uncertainty will occur when it was equally likely that a decision could be made at any level of the organization (a maximum dispersion of decision-making). Entropy calculated from Shannon's formula is increased when the uncertainty of a situation also increases (more dispersion of decision-making authority), attaining a maximum when all probabilities are equal. Entropy and dispersion may increase as the number of levels of the organization is raised. The difficulty with using Shannon's formula of course resides in the stygian problem of finding the probability that a decision is rendered at a given level. If everyone is independent, we can relate probabilities to frequencies — how often decision are made at each level and then use entropy as a measure of dispersion in the analysis of business and economic data. Hildenbrand and Paschen were early believers in entropy in the analyses of economic data, using it as a measure of industrial concentration, though Theil applied it more extensively to industry sales, foreign trade and corporation balance sheets. Thus should the calculations yield an entropy value for imports of leading commodities of 2.76 for 1950 and 2.86 for 1960, we may conclude that disorder in the distribution of imports increased during this decade The results of these calculations have been interpreted syncretically as signifying a freedom of choice, uncertainty, disorder, information content and information gains or losses. A popular use of entropy is in the analysis of market structure where in a monopoly situation the buyer has no choice as to the firm to be patronized, the lack of choice is reflected in the entropy measure (a value of zero for a perfect monopoly). Entropy is a measure of information in the sense that it focuses on the uncertainty surrounding a transmitted message. The problem of aggregation in accounting and in financial analysis is primarily that of providing the user with sufficient insight into the financial condition of the firm or the market. Rearranging and combining items in various ways may reduce the user's insight and cause a different meaning to be attached to the message.

Davis and Lisimin suggested money as the analog of economic entropy. The economic progress is entropic; it neither creates nor consumes matter or energy but only transforms low-entropy systems to higher levels. Money constitutes the economic equivalent of low entropy. We can seek a conversion factor to find the entropy equivalent of economic value. Perhaps the earliest work in irreversible thermodynamics was Thompson's examination of the creation and maintenance of a concentration difference across a film of liquid with a temperature difference. The heat flow caused the concentration gradient; the concentration gradient implied order (a higher order than before and less entropy). All caused by the heat flow. Social organizations as diverse as informal friendships and cultures have definable conditions of order and efficiency just as surely as do steam engines. Three basic flow categories have been identified which cross corporate boundaries: matter, energy and information. Specific flows into the system include, among others, raw materials, supplies, equipment, services, work by employees, information and capital, usually expressed in monetary not entropy terms. The flow-outs of a manufacturing operation are products, by-products, waste, information and dividends. There are also pressures from the environment: government regulations and actions of competitors. Internal processes include the aging of employees and equipment, generation of novel information, changes in organizational structure and conflicts among employees. The search for a relationship between value and entropy began around 1880 when money became "socio-biological energy". Suppose we are to boil water using alternatively, wood, natural gas and coal as the fuel. The entropy change of the water, in going from liquid water to steam is the same if we begin at the same temperature and pressure and end similarly for each of the three different burning processes. But the cost of the product, steam, depends on which fuel is used and hence the entropy money conversion cost is not fixed. The efficiency of the burning process is also important. Entropy changes can be calculated for: water to ice transformations; carbon dioxide gas to solid dry ice changes; the graphite-diamond correspondence, and helium gas-helium liquid phase inversions. Depending on the energy, equipment and labor cost, we can produce empirical correlations to describe an entropy-money conversion factor. If these factors were readily available, we could make an entropy balance around a corporate open system and establish its entropy exchange with the environment. Higher entropy values are synonymous with a healthy level of competition (a more disorganized market with competition distributed).

Hahn attempted to establish a relationship between the usual corporate performance indicators such as profits, sales, etc., and informational entropy. The goal was to correlate fiscal performance with internal hierarchical structure (the table of organization). Corporate structure was written as blocks, such as those for the President, and the Vice President for Sales. Each block was characterized by a power factor containing two parameters: (1) the fraction of the total corporate budget controlled and (2) the hierarchical level below the President (where the top rung is 1, Vice Presidents are 2). Thus the measure of power each organizational block carries is directly related to its budgetary responsibility and inversely proportional to the distance from the ultimate decision maker, the President. Changes in fiscal performance were mapped along with entropies calculated by Shannon's formula. This preliminary study of Hahn's suggested that an optimum structural configuration may exist, where performance and productivity can be maximized within an organization by establishing which internal structure is most conducive for optimum information flow. Shannon's formula is a useful tool here for deriving entropy data because: (1) maximum entropy seems to correlate well with the most highly disordered state; (2) increasing the number of hierarchical blocks raises the organizational entropy; and (3) higher entropy values are synonymous with greater competition. An inefficient and disorganized company should evidence high entropy and high operating expenses. There can be, with some structures, too little informational entropy (the capacity to transmit information) where insufficient capacity to transmit information means too much information is being stored and not used. This excess of stored information (calculated from Shannon's formula) could likely garble the lines of communication between the various departments of a corporation, manifested by tedious administrative and clerical detail as well as excessive interdepartmental formality and reserve. The result could be a hierarchy so complex and rigid that the system's ability to disseminate information quickly to its components and to respond with flexibility in the face of change is seriously undermined.

CHAPTER 14

Understanding the Aging Corporation

Organizations can no longer be changed by imposing a model developed elsewhere.

The change process needs to make use of the corporation's nature, by recognizing tendencies and building upon them. We begin a change process by acknowledging the basic worth and self-esteem of the individuals within the corporation, investing them with value, noting their past accomplishments, and with this as a springboard, developing the changes needed.

How do we create structures that move with the change, that are flexible and adaptive, even boundary-less, that enable rather than constrain?

Change is the norm. Change or die. Our mantra should be: be prepared when beginning the change process. Remember that we need to attain a far-from-equilibrium state. No minor tinkering will do, for if we are in the near-equilibrium mode, it is too easy to return to the comfortable equilibrium condition and delay or stop the change process. (When a system is far-from-equilibrium, creative individuals have enormous impact.) In the far-from-equilibrium condition we are open to big changes, having arrived at a bifurcation point, far out, away from the comfort zone of equilibrium. At the bifurcation point there may be two options available: one path may lead to significant and invigorating changes; the other may lead to disaster. Obviously we need to be able to recognize these possibilities. And so we take the risks in order to find the path to a more desirable level on the corporation lifecycle.

Autopoiesis means self-reproduction.

One of the characteristics of a living system is the ability to continuously reproduce and renew itself and to regulate this process in such a way that the integrity of its structure is maintained. It would seem to be destructive for a corporation to emulate this behavior since the change process would then be neutralized by the autopoiesis.

Dissipative structures are open systems that have flows of material, energy, and information through their boundaries.

There is also inner turmoil within its dissipative borders. Corporations can evolve as dissipative structures, always maintaining that tension of life, the requirement that they need some resistance to work against, to be tested, and to be challenged on their assumptions.

The dangerous alternative is to seal the boundaries and prevent flow through them. Such a closed systems now contains the seeds of its own destruction, according to the second law of thermodynamics. Closed systems, without outside intervention, must wear down, become inefficient, accumulate poisonous waste products, and eventually, in too short a time, die.

Stability through fluctuations, for open systems, means that disorder and the dissipative process can be a source of a new order. Significant fluctuations (forces for change) drive us far-from-equilibrium, where many new options become available to choose. Though being in equilibrium feels more comfortable, it is the path we choose in the far-from-equilibrium state that feels orderly and stable.

Organizations are not machines, they are processes.

Some think that an organization can be defined and controlled by focusing on its individual components and discrete tasks.

However, this can produce a proliferation of separations, a reduction of the whole into its component parts and an erroneous picture of the organization. Instead, we might try to divide an organization into different groups of connections. Corporations organized around core competencies are stable, with well-defined boundaries and openness. You have in such corporations a portfolio of skills rather than a portfolio of business units.

Space is not empty, in the universe as well as in an organization.

There are invisible fields of energy and information, which fill the space. Vision and creativity can also be considered invisible fields.

The Heisenberg uncertainty principle applies everywhere in life.

This means that by examining closely a corporation or parts of it, we inevitably change the parts or the whole organization. We need to understand that.

Don't define a person in terms of his or her authority relationship.

Instead, conceptualize the pattern of energy, and information, which flows to and from this person. We are conduits and also a source of energy and information flows.

Entropy is the measure of disorder in a system.

Natural processes proceed with increasing entropy and hence increasing disorder: towards maximum entropy, maximum disorder, maximum randomness and total independence and unconnectiveness of its parts (a definition of a death). Death is the universal attractor. We all end up there, no matter where or when we start.

Entropy is not something tangible, not anything capable of being seen or touched. Hence, entropy is a very difficult concept to grasp. It is not a solid; it is not hot or cold and does not have a physical consequence, such as temperature. We can say that entropy is a measure of the disorder of a system and show that more disorder means higher entropy content. For example, a container is divided in two, with molecules A on one side of the partition and molecules B on the other. This system has a certain entropy content and is ordered in its own way. Remove the partition which divides the container and after a while molecules A and B will mix thoroughly. This mixed state has a higher entropy content and we say it is more disordered.

The automobile engine, before ignition, contains mixtures of gasoline and air under pressure, a relatively ordered system compared to the exhaust gases, after ignition, coming out of the tail pipe. We say the process of combustion has proceeded from a state of low entropy to one of high entropy (out of the tail pipe). Real processes tend to go in the direction of increasing entropy. Aging can be envisioned as an irreversible process of entropy accumulation. Getting older means having less control of body functions; being more disordered. Death is the ultimate disorder, a state of maximum entropy.

Entropy as a concept originated during the development of the scientific field of the thermodynamics, the study of how we can convert heat into work. It was defined by the German physicist, Clausius, in 1869. Earlier, in his studies of the efficiency of heat engines, Carnot in 1824 introduced the second law of thermodynamics, which essentially says that systems will run down of their own volition if left to themselves. In other words, the entropy content tends towards a maximum. This increasing entropy could be an indication of the direction in which the system is inclined to go.

Unless there is outside intervention, the second law of thermodynamics codifies the one-sidedness of time, or time's arrow. We can only move forward, that is, time is irreversible.

Everything we know is tending towards chaos (unless there is outside intervention), towards an equilibrium with the environment. From Von Bertanlanffy, the significance of the second law can be expressed in another way. It states that the general trend of events is directed toward states of maximum disorder, the higher forms of energy such as mechanical, chemical, and light energy being irreversibly degraded to heat, and heat gradients continually disappearing. Therefore, the universe approaches entropy death when all energy is converted into heat at low temperature, and the world process comes to an end.

It was not until the 1950's that entropy started to seriously emerge in discussion of living phenomena. Complicated biological processes such as cell differentiation, growth, aging, were now analyzed from the second law of thermodynamics and entropy calculations made. Bailey wrote that entropy is a very viable concept for the biological and social sciences. It applies to both open and closed system. It can be discussed in terms of organization or order. Jones wrote that one common feature of biological processes is their unidirectionality in time, that is, they are not reversible, except under special circumstances. Since entropy is the only physical variable in nature which generally seems to parallel the direction and irreversibility of time, these should be fertile areas for the effective use of entropic models.

Much of the historical development of entropy had dealt with isolated or closed systems. A closed system is one which cannot exchange energy or matter with the surrounding environment. The second law of thermodynamics states that a closed system must evolve to a state in equilibrium with its environment, a condition of maximum entropy. Open systems are those which can exchange both matter and energy with the surroundings. Obviously we, the living, are examples of open systems. Open systems must maintain the exchange of energy and matter in order to sustain themselves, or slow the approach to the final state, death.

We can say that entropy accumulation within the living system is composed of two parts, one being the internal entropy production based on the myriad of irreversible chemical reactions which constitute the chemistry of life. Secondly, there is the entropy flow through our boundaries, such as in the food consumed, air inhaled and exhaled, biological waste products, heat loss from the skin, etc. The internal entropy production derived from our chemical reactions always proceeds in the direction of increasing our entropy content, since the chemistry of life in inherently irreversible. However, the food in, air in and out, waste products out, and heat out may in total contribute positive or negative entropy flows through our boundaries, which then affect the rate of accumulation of entropy in our living body.

Old age or senescence may be the decline in our ability to produce free energy. Less free energy means diminished cell function. Vitality might be defined as our biological and thermodynamic strength, the ability to expend energy to restore ourselves to near original conditions. In other words, aging can be thought of as the process of loss of vitality.

Zotin proposed that we evolve towards a final state, death, by a series of changes, each change called a stationary state. We settle into a stationary state, stay for a while, until pushed to the next, and the next. This is clearly seen in the transformation of butterfly larvae and pupae, dramatic physical changes. Balmer applied entropy concepts to the study of aging annual fish. This species displays all the characteristics of birth, growth, aging, and senile death, over a short twelve-month period. Balmer identified entropy flows into and out of the fish, such as food, excrement and body heat dissipation.

Positive feedback is the source of instability.

When a plus output signal is returned (fed back) to the plus input signal, a larger plus, or positive output results. Tapping off part of this new plus output and again adding it to the plus input continues the process of a pressure build up with time in the output. Eventually the system output is driven far from its stationary state and we may be on the verge of disaster (we hope not) or on the verge of something new, (something innovative or creative). Positive feedback drives self-organizing systems to change.

Self-organizing systems find ways of adjusting to the external influences.

Self-organizing, self-referencing organizations, upon finding new information, can develop new structures and new missions to accommodate to these new initial conditions, which affect their chemistry. The more freedom there is to be self-organizing, the more order exists. Self-organization means having the freedom to grow and evolve. A self-organizing system must remain consistent with itself and its past. It is globally stable while making internal changes.

The Belousov-Zhabotinsky chemical reaction is a dramatic example of this, where, by continuously adding new amounts of reactants to a chemical reactor which contains certain chemicals, many interesting intermediate chemical compounds form in the reactor, live for a while and then disappear, to be replaced by new compounds. The result in the glass reactor is a beautiful, colorful display of ever-changing patterns and colors which disappear and reappear as long as more reactant chemicals are added and chemical products drawn off.

Scientists build explanatory structures, telling stories, which are tested to see if they are stories about real life.

Surprise is the major route to discovery. Being willing to be embarrassed and still carry on is a characteristic of creativity. Creativity is staying comfortable with uncertainty. Great innovations may at first appear muddled and strange, only partly understood by their discoverer.

Our universe is not merely a universe of things, but also a universe of a nothingness we call information.

Aging theories

Aging may be the evolution towards a more probable state, the equilibrium state. Schrodinger wrote that living systems survive by avoiding the rapid decay into the inert state of equilibrium. Jones proposed that the approach to equilibrium is a sign of death. Death may also be thought of as the attaining of a critical, maximum state of entropy during our journey towards equilibrium. Schrodinger further stated that a living organism continually increases its entropy... and tends to approach the dangerous state of maximum entropy, which is death.

Death can be when a critical amount of randomness is attained, when a certain amount of disorganization is suffered. Thus aging is a randomizing process, a disorganizing process. In terms of stability theory, equilibrium is the point or region of attraction. We are drawn relentlessly towards equilibrium. Life may be considered analogous to the spring-wound watch, where the timepiece may stop by one of two possible mechanisms. It can simply wind down (approach equilibrium), or the internal mechanism can somehow fail and the watch dies prematurely. Thus we can describe equilibrium deaths and instability or catastrophic deaths.

Life may be considered a temporary upset from, or perturbation of, equilibrium. Equilibrium is absolutely stable, a universal attractor. Equilibrium always wins. Evolution may be the natural process of prolonging the time that an organism spends in the far-from-equilibrium state. Genetics still plays its role in determining longevity potential, but is not in conflict with the ideas presented here. The tendency to return to equilibrium will always apply. Death will always have a probability of absolute certainty.

Aging People versus the Aging Corporation

People age because the structural material such as collagen in our bodies becomes dried out, more rigid, and cross-linked and resistant. It is through these colagenous membranes that enzymes, hormones, oxygen, and the other vital metabolic chemicals diffuse into our cells. There they allow our cells to function vigorously and properly.

Corporations age because the boundaries between units become closed and information and energy cannot efficiently flow from one part of the organization to the next.

People age because our somatic cells cause deleterious mutations and become defective.

Corporations age because their units become inefficient and defective.

People age because our immune system begins to attack our own cells, thinking they are foreign. Our body literally devours itself through ill design and ill function.

Corporations age because some their units begin attacking others, rationalizing that these units are foreign to and dangerous for, the mission of the organization, and need to be neutralized.

People age because free radicals disruptively enter into the chemistry of the cells and cause poisonous chemical "garbage" to form in the cells.

Corporations age because disruptive employees undermine the chemistry of the units.

People age because skeletal muscle shows marked changes in ability and strength.

Corporations age because they lose vitality, resolve, and strength to carry out their mission and compete effectively.

People age because there are lipofuscin (old age pigments) in the heart, causing the heart to have difficulty contracting and expanding.

Corporations age because some of their people seem to accumulate "rust" as they do their dead end jobs.

People age because of wear and tear in their body parts.

Corporations age because employees wear out.

People age because of an accumulation of toxic waste products.

Corporations age because of an accumulation of the toxicity of bad employee morale.

People age because of cell senescence, which is caused by the loss of irreplaceable enzymes.

Corporations age because key people leave.

People age because of an accumulation of entropy and disorder in their bodies.

Corporations age because of the same reason.

People age because of a loss of vitality, the energy level necessary to survive.

Corporations age because of the same reason.

People age because they cannot muster enough energy to meet a short-term challenge, and they die.

Corporations age because they behave like old people and then die.

People age because some irreplaceable cells like brain cells diminish in number.

Corporations age because of the loss of smart people.

People age because their energy metabolism sinks below the critical level needed to maintain life.

Corporations age because of the same reason.

People age because mutations in the cells cause errors of replication.

Corporations age because some units begin new activities without authorization, and perform poorly.

CHAPTER 15

Hershey Corporate Lifecycle Assessment and Entropy Analysis

Definitions

ai: A major activity that a unit does. Scaled 5, 3, 1, 0, as very important, important, not very important, and the unit does not do this.

$\sum$***ai:*** The numerical sum of the major activities a unit does.

bi: Fractional Budget. The budget of a unit divided by the total budget for the organization being studied.

Pi: The power of a unit. $Pi = bi/Si$, where Si is the shell (level) that the unit occupies in the organizational structure.

σPi: The sum of power of all the units that one unit controls. This is obtained by following the boss lines connecting the units.

Boss Relationship: The boss for a unit is defined to be at least one shell above it, with whom it shares a major portion of its activities.

Methods I and II: Procedures for deciding how to fill the shells in an organization. Based on principles which determine how electron shells are filled in an atomic structure. Method I requires each shell to have the same total activity content. Method II requires that a unit can only be a member of a particular shell if its $\sum ai$ is within the activity range for the shell.

Ci: The cumulative power of a unit. This incorporates its own power and the power of the units it controls. $Ci = Pi + \sigma Pi$.

fi: Fractional cumulative power. $fi = Ci/\sum Ci$ where $\sum Ci$ is the sum of the cumulative power values of all the units in the organization.

The Six Hershey Parameters: Scaled from 0.00 (Best) to 1.00 (Worst)

SR-A: Structural Redundancy-Activities. The degree of overlapping activities in the organization.

SR-G: Structural Redundancy-Geometry. A measure of the number of shells (levels) in the organization.

SR-I: Structural Redundancy-Interactions. The amount of shared activities for the units that bypass the bosses.

PD-CG: Power Distribution-Center of Gravity. Gives the location of the center of power in the organization.

PD-SN: Power Distribution-Symmetry Number. A measure of how the power is distributed within the organization.

ST/STmax: Total (T) Entropy (S) divided by the Maximum (max) Total (T) Entropy (S). This measures the efficiency of information flow.

Significance of the Six Hershey Parameters

Parameter	Best Case 0.0 for each parameter	Worst Case 1.0 for each parameter
SR-A (Structural Redundancy-Activities)	No overlapping activities	All units totally independent and doing the same activities.
SR-G (Structural Redundancy-Geometry)	Two levels in the table of organization	Totally vertical table of organization.
SR-I (Structural Redundancy-Interactions)	No shadow organizations	All units bypass their bosses and are totally independent.
PD-CG (Power Distribution-Center of Gravity)	Power at the top	Power shifted away from the top unit. The lower units are powerful.
PD-SN (Power Distribution-Symmetry Of Power Distribution)	One unit most powerful. Clusters of power.	Power distributed evenly. No tension between units. All independent.
ST/ST max (Informational Entropy)	Perfect information flow	No information lines between units. No bosses. No responsibilities. All units overlap in activities. Maximum dispersion, maximum randomness, and maximum disorder.

Hershey Corporate Lifecycles

	Human Age Year (% of 100)	Hershey Age Scale Based on % of 1.0 for each parameter	Hershey Age Scale Cumulative for the 6 parameters (% of 6.0)
Death	100	1.00	6.00
Moribund	90–100 (10%)	0.90–1.00 (10%)	5.40–6.00 (10%)
Old-Bureaucratic	70–90 (20%)	0.70–0.90 (20%)	4.20–5.40 (20%)
Mature-Stable	50–70 (20%)	0.50–0.70 (20%)	3.00–4.20 (20%)
Prime	30–50 (20%)	0.30–0.50 (20%)	1.80–3.00 (20%)
Youthful	0–30 (30%)	0.00–0.30 (30%)	0.00–1.80 (30%)
Birth	0	0.00	0.00

Scale 1.00		Scale 6.00
1.00	**Death**	6.00
0.90	**Moribund**	5.40
0.90	**Old-Bureaucratic**	5.40
0.70		4.20
0.70	**Mature-Stable**	4.20
0.50		3.00
0.50	**Prime**	3.00
0.30		1.80
0.30	**Youthful**	1.80
0.00	**Birth**	0.00

Calculations Involving the Six Hershey Parameters

Example

A matrix can be developed for an organization, as shown below, where $x1, x2 \ldots$ are the names of the various units, and $a1, a2 \ldots$ are major activities for the organization, and 0 and 1 represents the "no" and "yes" answer to the question as to whether a unit does or does not do that activity. In a more sophisticated approach to be shown later, the ai are values of 5, 3, 1, 0, where 5 denotes very important and 1 is not important.

Table I

	a1	a2	a3	a4	a5	a6	a7	a8	a9	$\sum ai$	Li
x1	1	1	1	1	1	1	1	1	1	9	1
x2	0	0	1	0	0	1	0	0	1	3	4
x3	1	0	0	1	0	0	1	0	0	3	4
x4	1	0	1	0	1	0	1	0	1	5	2
x5	0	1	0	1	0	1	0	1	0	4	3

$$GS = 24$$

where

$\sum ai =$ the horizontal sum of the activities for unit, i.

$GS =$ the vertical sum of $\sum ai$ for all units. $GS = 24$ includes all the overlapping activities. In this example, $x1$ and $x5$ share $a2$.

$Li =$ the level of unit, i, in the structure. Li is determined from the magnitude of $\sum ai$.

$x1$ has $\sum ai = 9$ and is positioned as level 1, $x4$, with $\sum ai = 5$ is at level 2, etc. $T =$ the minimum number of horizontal defining activities. In this example, $T = a1 + a2 + \cdots + a9 = 9$. In a more sophisticated approach to the shown later, $T = $ (the sum of the ai) multiplied by the average weighting factor. Therefore, $T = (9)(3) = 27$. $n =$ number of units in the organization.

Table II

Unit	Level	Bi	Pi $Pi = bi/Li$	σPi	Ci $(Pi + \sigma Pi)$	Fi $Ci / \sum ai$
x1	1	0.1	0.10	.275 (0.025 + .10 +.05 + .10)	.375 (.10 + .275)	.484
x2	4	0.1	0.025	0	.025	.0322
x3	4	0.4	0.10	0	.10	.129
x4	2	0.1	0.05	0	.05	.0645
x5	3	0.3	0.10	.125	0.225 (.10 + .125)	.290
			$\sum\limits_{i}^{n} Pi = .375$		$\sum\limits_{i}^{n} Ci = .775$	$\sum\limits_{i}^{n} fi = 1.00$

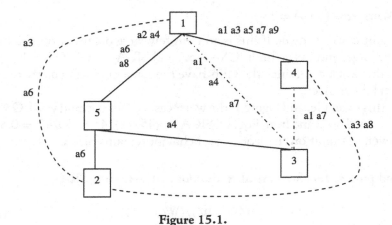

Figure 15.1.

Notice in Figure 15.1: solid lines denote boss lines and dotted lines represent other shared, non-boss interactions. Notice also there are no lines connecting $x4$ and $x5$ since they share no ai.

Six Hershey Parameters for Characterizing the Efficiency of the Structure

First parameter: Structural Redundancy-Activities (SR-A)

$$(SR\text{-}A) = (GS - T)/GS, \tag{1}$$

where, $GS - T =$ the difference between the vertical sum of $\sum ai$ for all units in Table I, minus the minimum number of defining activities. $GS - T$ represents the overlapping of responsibilities within the organization. It can be considered a measure of the inefficiency within the structure. In the Example (Table I) and Figure 1, $GS = 24$, $T = 9$, and SR-A $=$ $(24 - 9)/24 = 15/24 = 0.625$. SR-A is confined between a best case (SR-A $= 0$) and a worst case (SR-A $= 1$) by considering these limiting scenarios.

(a) *Best case (SR-A $= 0$)*

Each unit does its own unique job, its own ai, with no overlapping of responsibilities among the units. In this best case, $GS = T$, and from Eq. (1), SR-A $= 0/GS = 0$.

(b) *Worst case (SR-A = 1)*

Each unit does all the *ai*. If this is an infinitely large organization, GS can be very large, and in the limit, GS → ∞, and SR-A → GS/GS = 1.

In this worst case, since the units have the same number of *ai*, there are no level differences.

In this Example, and Figure 1, the worst case scenario would yield. GS = $5 \times 9 = 45$, $T = 9$ and from Eq. (1), SR-A = $(45-9)/45 = 36/45 = 0.80$. However, the limit on the worst case behavior remains SR-A $= 1$.

Second parameter: Structural Redundancy-Geometry (SR-G)

$$\text{SR-G} = (WO - W)/(WO - W_{max}), \qquad (2)$$

where
   W = average number of units on a level = n/L
   n = number of units in the structure
   L = number of levels in the structure
WO = best case
W_{max} = worst case

SR-G can be seen from Eq. (2) to range from the best case, when $W = WO$ and SR-G $= 0/(WO - W_{max}) = 0$, to the worst case when $W = W_{max}$, and SR-G = $(WO - W_{max})/(WO - W_{max}) = 1$.

(a) *Best case (SR-G = 0)*

All things being equal, the best structure is as shown below. Notice in this best case, information flow is relatively easily accomplished from bottom to top. Communications are simple. For this best case, $n = 5$, $L = 2$, and $W = WO = 5/2 = 2.5$.

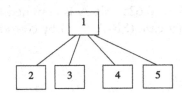

(b) *Worst case (SR-G = 1)*

Again, all things being equal, the worst structure can be as given below.

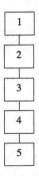

In this worst case, information flow needs to move through many layers before reaching the top. Distortion and loss of information are significant. For this worst case, $n = 5$, $L = 5$, and $W_{max} = 5/5 = 1.0$.

In this Example (Table 1) and Figure 1, $n = 5$, $L = 4$, $W = 5/4 = 1.25$, and from Eq. (2), SR-G $= (2.5 - 1.25)/(2.5 - 1.0) = 0.833$.

Third parameter: Structural Redundancy-Interaction (SR-I)

The dotted lines in Figure 1 indicated interactions between units. These interactions complicate the flow of information among the units which are linked by solid (boss) lines. The interactions are a way of bypassing the boss. Thus the dotted lines describe confusion or disorder in the information flow. This inefficiency, calculated by the methods of the next section, finds expression in the third structural redundancy, SR-I, and Eq. (3)

$$SR\text{-}I - SI/ST, \qquad (3)$$

where

S = informational entropy, a measure of the disorder within the structure. More details are given later

SI = Informational entropy resulting from the interactions (the dotted lines).

$ST = S + SI$ = total informational entropy within the structure.

SR-I can be seen from Eq. (3) to be defined between the best case (no interactions, $SI = 0$, and SR-I $= 0$) and the worst case (overwhelming interactions, $SI \to \infty$ and SR-I $\to SI/SI = 1$).

Fourth parameter: Power Distribution-Center of Gravity (PD-CG)

Equation 4 defines the Center of Gravity (CG)

$$CG = \frac{[1x\sigma P(\text{Level1}) + 2x\sigma P(\text{Level2}) + \cdots]/n}{<P>}, \tag{4}$$

where

$\sigma P(\text{Level 1}) = $ sum of the power of all units on Level 1
$\sigma P(\text{Level 2}) = $ sum of the power of all units on Level 2
$\quad\quad <P> = $ Average power $= \sum_i^n Pi/n$
$\quad \sum_{i=1}^n Pi = $ sum of the power of all units

From the definition of power, $Pi = bi/Li$ and $<P>$, Eq. (4) can be rewritten as

$$CG = \frac{\left[1x\sum_{i=1}^n \left(\frac{bi(\text{Level1})}{1}\right) + 2x\sum_{i=1}^n \left(\frac{bi(\text{Level2})}{2}\right) + \cdots\right]/n}{\sum_{i=1}^n Pi/n}$$

$$= \sum_{i=1}^n bi \Big/ \sum_{i=1}^n Pi$$

With $\sum_{i=1}^n bi = 1$ (by definition), Eq. (4) becomes Eq. (5).

$$CG = \frac{1}{\sum_{i=1}^n Pi}. \tag{5}$$

Thus, for structures with all the power at the top, $P_1 = b_1/L_1 = 1.0/1 = 1.0$ and $\sum_2^5 Pi = 0$ which yields in Eq. (5), $CG = 1.0$. Alternatively, with the power focused at the bottom, $P_5 = b_5/L_5 = 1.0/5 = 0.2$, $\sum_1^5 Pi = 0.2$ and $CG = 1/0.2 = 5$. From Table II and Figure 1, Eq. (5) yields $CG = 1/0.375 = 2.667$.

All things being equal, the level of the Center of Gravity (CG) significantly changes the flow of information. Thus entropy, S, is dependent on the location of the CG. The parameter which reflects this is given by

Eq. (6), the Power Distribution-Center of Gravity, (PD-CG), the fourth parameter.

$$PD\text{-}CG = \frac{CG - CG0}{CG_{max} - CG0},\qquad(6)$$

where

$CG0 =$ the center of gravity for the case, where all the power is focused at Level 1 (the President). It can be shown that when this is the case, entropy, S, becomes very small, approaching $S = 0$. Therefore, $CG0 = 1.0$.

$CG_{max} =$ the center of gravity for the worst case, where all the power resides on the lowest level of the structure. It can be shown that when this is true, entropy, S, becomes very large, approaching S_{max}. Therefore, $CG_{max} = L$, the number of levels in the structure.

When $CG = CG0$, PD-CG $= 0$ (best case) and when $CG = CG_{max}$, PD-CG $= 1$ (worst case).

In the Example, Table II and Figure 1, and with Eq. (6), $PD - CG = (2.667 - 1)/(4 - 1) = 1.667/3 = 0.556$.

Fifth parameter: Power Distribution-Symmetry Number (PD-SN)

All things being equal, the deviation of fi from its mean value $<fi>$ significantly affects the flow of information. In other words, the more the fi in the structure are the same, the more disordered the information flow. The Symmetry Number (SN), defined by Eq. (7), allows the calculation of these deviations.

$$SN = \sum_{i=1}^{n} (fi - <fi>)^2/n,\qquad(7)$$

where $<fi> = \sum_{i=1}^{n} fi/n = 1/n =$ mean value of fi.
   Note: $\sum_{i=1}^{n} fi = 1$

(a) Best case (SN0)

When the power is concentrated at the top of the organization, information flow tends to be optimized, and there is less disorder in the structure (S is minimized). This means, $f1 = 1$, and $f2 = f3 = fn = 0$.

From Eq. (7), SN0 can be determined, to yield Eq. (8).

$$SN0 = [(1 - 1/n)^2 + (n - 1)(- 1/n)^2]/n$$
$$= [1 + (1/n)^2 - 2/n + (n - 1)(- 1/n)^2]/n$$
$$= [1/n^2[(n - 1) + 1] + 1 - 2/n]/n$$
$$= \left(\frac{1}{n^2}n + 1 - 2/n\right)\Big/n$$
$$= (1/n + 1 - 2/n)\Big/n$$
$$= (1 - 1/n)/n$$

and finally,

$$SN0 = (n - 1)/n^2. \qquad (8)$$

(b) Worst case (SN$_{max}$)

When $fi = <fi>$, $SN = SN_{max} = 0$ from Eq. (7).

In the Example Table II and Figure 1, and with Eq. (7), the Symmetry Number (SN), can be calculated. Also, from Eq. (8), the best case SN0 is determined. The worst case symmetry number SN$_{max}$, is always zero. The calculations are shown below.

SN (from Eq. (7))
$$= [(.484 - .2)^2 + (.0322 - .2)^2 + (.129 - .2)^2$$
$$+ (.0645 - .2)^2 + (.29 - .2)^2]/5$$
$$= [(.284)^2 + (- .168)^2 + (- .071)^2 + (- .136)^2 + (.09)^2]/5$$
$$= (.0806 + .0282 + .00504 + .0185 + .0081)/5$$
$$= 0.1404/5$$

And finally, SN = 0.02808
Also, SN0 (from Eq. 8) = $(5 - 1)/5^2 = 0.16$

The fifth parameter, Power Distribution-Symmetry Number (PD-SN), given by Eq. (9), is normalized, to be contained between 0 (the best case where SN = SN0) and 1.0 (the worst case, SN = SN$_{max}$ = 0).

$$PD\text{-}SN = 1 - SN/SN0. \qquad (9)$$

From Table II, and Eq. (9),

$$PD\text{-}SN = 1 - (.02808/16) = (1 - .1755) = .8245$$

Sixth parameter: informational entropy

Shannon's informational entropy formula, applied originally to communications theory, has been adapted for the business world, as well as in the study of gene information content, and a host of other applications. In all of these cases, high entropy is associated with much disorder, randomness, and more error possibilities.

Shannon's informational entropy formula is usually written in terms of logarithms to the base 2 ($\log_2 x$), but here, natural logarithms ($\ln x$) will be used. Natural logarithms are more common, and since entropy ratios will be calculated, it makes sense to do this. (The conversion factor between natural and base 2 logarithms is given at the end of this section.) Also, for simplicity, the minus signs arising from $\ln x$, when x is less than 1.0, will be ignored. Equation (10), a modified version of Shannon's informational entropy formula, is the starting point for entropy calculations.

$$S = \sum_{i=1}^{n} f_i \ln f_i, \tag{10}$$

where $\sum_{i=1}^{n}$ = the sum of all n units in the organization.

If C_i in the Example Table II are equal, then $\sum C_i = nC_i$, and $f_i = C_i / \sum C_i = C_i / nC_i = 1/n$.

In other words, to get S_{max}, use $f_i = 1/n$.

In Eq. (10), this yields

$$S = \sum_{i=1}^{n} \frac{1}{n} \ln \frac{1}{n} = \sum_{i=1}^{n} \frac{1}{n} (-\ln\ n) = n[1/n(-\ln n)] = -\ln n = \ln n.$$

(if we ignore the minus sign).

Shannon's formula yields S_{max} when the f_i are equal (the worst case scenario). From the proceeding development, this becomes Eq. (11):

$$S_{max} = \ln n. \tag{11}$$

In the example, Table II and Figure 1, using Eqs. (10) and (11), S and S_{max} can be calculated.

$$S = .484 \ (\ln .484) + .0322 \ (\ln .0322) + .129 \ (\ln .129)$$
$$+ .0645 \ (\ln .0645) + .290 \ (\ln .290)$$

$$= .484 \, (.726) + .0322 \, (3.436) + .129 \, (2.048)$$
$$+ .0645 \, (2.741) + .290 \, (1.238)$$
$$= .351 + .111 + .264 + .177 + .359$$
$$= 1.262$$

Therefore,

$$S = 1.262$$

and

$$S_{max} = \ln 5$$

or

$$S_{max} = 1.61$$

and

$$S/S_{max} = \text{fractional distance from the worst case}$$
$$= 1.262/1.61 = 0.78$$

When $S = S_{max}$, there is no stored information. It is all potential. When $S = 0$, $S/S_{max} = 0$, which signifies all stored information. It is the ratio, S/S_{max} which is one form of the sixth parameter: informational entropy.

Interactions Between Units (Dotted Lines) and the Effect on Informational Entropy

Interactions (non-boss sharing of ai) diffuse the orderly flow of information, and hence generate disorder (and increased entropy). In Eq. (3), the entropy contribution generated by interactions, SI, was introduced. In calculating SI, it is hypothesized that the interacting units behave as doublets. This means they act as if they were the only units in the organization. As units within the doublets, they carry their own fiI, where fiI refers to the unit, i, interacting (I) within a doublet. This idea is perhaps best clarified with a specific calculation. From Example (Figure 1), the interactions are seen to be between units $(x1 - x2)$, $(x1 - x3)$, $(x2 - x4)$ and $(x3 - x4)$. Table III below illustrates the calculations and the meaning of fiI. The fi for xi are obtained from Table II.

Table III (Interactions)

Doublet	as interacting doublets
$(x1 - x2)$	
$f1 = .484$	$f1I = .484/.516 = .938$
$f2 = .0322$	$f2I = 1 - .938 = .062$
$\overline{.516}$	
$(x1 - x3)$	
$f1 = .484$	$f1I = .484/.613 = .790$
$f3 = .129$	$f3I = 1 - .790 = .210$
$\overline{.613}$	
$(x2 - x4)$	
$f2 = .0322$	$f2I = .0322/.0967 = 0.333$
$f4 = .0645$	$f4I = 1 - 0.333 = .667$
$\overline{.0967}$	
$(x3 - x4)$	
$f3 = .129$	$f3I = .129/1.94 = 0.665$
$f4 = .0645$	$f4I = 1 - .665 = .335$
$\overline{.199}$	

Now *SI* can be calculated, using Shannon's entropy formula, Eq. (10), modified to indicate the fact that the units are doublets. This means *fiI* values are to be used. However, it needs to be recognized that the entropy values obtained ought to reflect the reality, that though these units are interacting as doublets, seemingly alone in the organization, when in fact they are not alone. Therefore, to place unit *i* properly within the context of the organization as a whole, the actual *fi* will be used as a weighting factor. In other words, when calculating *SI*, use Eq. (12) for each unit within a doublet.

$$SI = fi[fiI \ln fiI] \text{ (per unit)}, \tag{12}$$

where fi = the weighting factor for unit *i* in the doublet.

From Table III, and Figure 1, the entropy of interactions (per doublet) can be calculated as shown.

$$
\begin{aligned}
SI\,(x1 - x2) &= f1\,[f1I \ln f1I] + f2\,[f2I \ln f2I] \\
&= .484\,[.938 \ln .938] + .0322\,[0.0622 \ln .0622] \\
&= .484\,[.938(.0640)] + .0322\,[.0622(2.781)] \\
&= .484\,(.0600) + .0322\,(.172)
\end{aligned}
$$

$$= .0290 + .00554$$
$$= 0.0345$$
$$SI\,(x1 - x3) = .484\,[.790\ln.790] + .129\,[.210\ln.210]$$
$$= .484\,[.790(.236)] + .129\,[.210(1.561)]$$
$$= .484\,(.1860) + .129\,(.328)$$
$$= .0900 + .0423$$
$$= 0.132$$
$$SI\,(x2 - x4) = .0322\,[.333\ln.333] + .0645\,[.667\ln.667]$$
$$= .0322\,[.333\,(1.100)] + .0645\,[.667\,(.405)]$$
$$= .0322\,(.366) + 0.645\,(.270)$$
$$= .01180 + .0174$$
$$= 0.0292$$
$$SI\,(x3 - x4) = .129\,[.665\ln.665] + .0645\,[.335\ln.335]$$
$$= .129\,[.665(.408)] + .0645\,[.335(1.094)]$$
$$= .129\,(.271) + .0645\,(.366)$$
$$= .0350 + .0236$$
$$= 0.0586$$

or

$$SI = .0345 + .132 + .0292 + .0586$$

and finally,

$$SI = 0.254$$

This augmentation of the structural entropy, S, (without interactions), Eq. (10), by SI, the entropy contribution arising from interactions, Eq. (12), produces the total entropy of the structure, ST, given by Eq. (13).

$$ST = S + SI. \tag{13}$$

From Eq. (10), the Example entropy was calculated to be $S = 1.262$, and from Eq. (12), $SI = 0.254$. This yields from Eq. (13), $ST = 1.262 + .254 = 1.516$, and the third parameter, Structural Redundancy-Interactions (SR-I),

$$SR\text{-}I = 0.254/1.516 = 0.168 \text{ from Eq. (3).}$$

Previously, one form of the sixth parameter, entropy, $S/S_{max} = 1.262/1.61 = 0.78$ was calculated. This is the entropy ratio for the structure, assuming no interactions. Now we seek an analogous ratio which will allow the more general case, that there are interactions. We already have ST, the entropy content for the structure with interactions. What remains is to develop a relationship, ST_{max}. This could be written as follows:

$$ST_{max} = S_{max} + SI_{max}, \tag{14}$$

where $SI_{max} = $ the entropy contribution to maximum entropy (S_{max}), arising from interactions.

Equation (11) showed $S_{max} = \ln n$, when all the fi are equal, and $fi = 1/n$. To get SI_{max}, the units are again considered to be acting as doublets, and hence the computation is similar to that shown in Table III, only this time, $fi = 1/n$. Table IV below is typical for all SI_{max} interactions. It shows one doublet. Recognize that the other doublets yield similar results.

Table IV (Interaction from the Example to get SI_{max})

Interactions doublet

As doublet

$(x_1 - x_2)$
$$f1_{max} = 1/n \quad f1_{max}I = \frac{1/n}{2/n} = 1/2$$

$$f2_{max} = 1/n \quad f2_{max}I = \frac{1/n}{2/n} = 1/2$$

$$\overline{2/n}$$

Therefore, per doublet interaction, from Eq. (12), and with fi_{max} as the weighting factor; using Table IV, SI_{max} can be determined.

Per doublet interaction:

$$SI_{max} = fi_{max}[fi_{max}I \ln fi_{max}I] + fi_{max}[fi_{max}I \ln fi_{max}I]$$
$$= 2\{fi_{max}[fi_{max}I \ln fi_{max}I]\}$$

Note in the above equation, there are 2 units per doublet interaction. SI_{max} becomes finally from Table IV.

Per doublet interaction:

$$SI_{max} = 2\{1/n[1/2 \ln 1/2]\} = 1/n \ln 1/2$$

or

$$SI_{max} = 1/n \ln 2 \text{ (if we ignore minus signs)}, \tag{15}$$

where

n = number of units in the organization.

If there are K doublets (K dotted lines), then the contribution to the maximum entropy by the interactions, Eq. (15), is SI_{max}, given by Eq. (16),

$$SI_{max} = K(1/n \ln 2). \tag{16}$$

The total maximum entropy of the structure (ST_{max}) allowing for interactions are found from Eqs. (11), (14) and (16).

$$ST_{max} = \ln n + K(1/n \ln 2)$$

or

$$ST_{max} = \ln (2^{k/n} n). \tag{17}$$

In the Example, with $k = 4$, and $n = 5$, ST_{max} becomes from Eq. (17)

$$\begin{aligned} ST_{max} &= \ln (2^{4/5} \times 5) = \ln (2^{0.8} \times 5) \\ &= \ln (1.741 \times 5) = \ln (8.70) \\ &= 2.163 \end{aligned}$$

From Eq. (13), $ST = 1.516$ and from Eq. (17), $ST_{max} = 2.163$, which yields:

$$ST/ST_{max} = 1.516/2.163 = 0.701,$$

the sixth parameter: entropy.

These results were obtained:

S	1.262	Entropy
SI	0.254	Entropy of interactions
ST	1.516	Total entropy
ST/ST_{max}	0.701	Total entropymaximum total entropy
SR-A	0.625	Structural redundancy-activities
SR-G	0.833	Structural redundancy-geometry
SR-I	0.168	Structural redundancy-interactions
PD-CG	0.556	Power distribution-center of gravity
PD-SN	0.824	Power distribution-symmetry

Some Mathematical Clarifications

1. If $fi = 0$, then $S = \sum fi \ln fi$ is an indeterminate form since $fi \ln fi = 0 \times \infty$. We can use L'Hopital's rule to evaluate this indeterminate form. As an illustration, use only one term in the S equation, as follows:

$$\lim_{fi \to 0} S = \lim_{fi \to 0} fi \, (\ln fi)$$

or

$$\lim_{fi \to 0} S = \lim_{fi \to 0} \frac{\ln fi}{1/fi} = \lim_{fi \to 0} \frac{1/fi}{-1/fi^2} = \lim_{fi \to 0} (-fi) = 0$$

Thus the contribution to S, in $\sum_{i=1}^{n} fi \ln fi$, is zero for each fi term that is zero.

2. In the calculations, where natural logarithms $(\ln x)$ are used instead of logarithms to the base 2, $(\log_2 x)$, the correction factor can be found as follows, where $\ln x$ means log to the base e

$$\log_2 5 = z \quad \ln 5 = x$$
$$5 = 2^z \quad\quad 5 = e^x$$

Therefore $2^z = e^x$ or

$$(z) \ln 2 = (x)$$

and

$$(\log_2 5)(\ln 2) = (x) = (\ln 5)$$

This gives

$$(\log_2 5)(.693) = \ln 5$$

or

$$(\log_2 5) = 1/.695 \ln 5 = 1.44 \ln 5$$

or, in general,

$$\log_2 N = 1.44 \ln N$$

Theoretical Basis for Determining the Bosses in a Corporate Hierarchy

The president is Unit 1 alone at the top of the table of organization (Level or Shell 1).

The president's ai are all 5 when the "weight factor" for ai is 5, 3, 1, and 5 is very important and 1 is not important.

The president is involved in all the ai. This means the president's $\sum ai$ is always given by $\sum ai = 5 \times$ (the number of ai for the organization). If the number of ai for the organization is 10, then the Presidents $\sum ai = 5 \times 10 = 50$.

The boss for the units on Level (Shell) 2 is Unit 1.

The boss of a unit is one level (shell) above, with the greatest value for (the sum of shared ai numbers) $\times$ (Power, Pi, of the possible boss). See the illustration below.

If the boss of a unit cannot be found on the level (shell) above, proceed to the next level (shell) above.

Except under unusual circumstances, Unit 1 cannot be the boss for units on Levels (Shells) 3, 4, 5, etc.

For example, the boss of Unit 2 might be Units 3, 4 or 6.

| 2 | a4, a6 | | 3 |
| | 3 + 3 = 6 | | |

Boss Criteria

(6) (P3) = (6)(0.023) = 0.14

$a_4 = 3$
$a_6 = 3$
$P3 = .023$

| 2 | a1, a4, a6 | | 4 |
| | 5 + 3 + 3 = 11 | | |

Boss Criteria

(11) (P4) = (11)(0.087) = 0.96

$a_1 = 5$
$a_4 = 3$
$a_6 = 3$
$P4 = .087$

| 2 | a1 | | 6 |
| | 5 | | |

Boss Criteria

(5) (P6) = (5)(0.029) = 0.14

$a_1 = 5$
$P6 = .029$

Therefore the boss for [2] is [4]

Note: The numerical values for the ai are those for Unit 2.

Theoretical Basis for Determining the Summation of Power (σPi)

The σPi for a unit is the power (Pi) it controls (not including its own Pi). This is determined by following the boss lines (the solid lines) in the hierarchy.

For the simple structure shown here

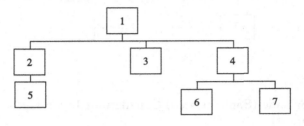

$$\sigma P1 = (P2 + P5) + (P3) + [(P4) + (P6 + P7)]$$
$$\sigma P2 = P5$$
$$\sigma P3 = 0$$
$$\sigma P4 = (P6 + P7)$$
$$\sigma P5 = 0$$
$$\sigma P6 = 0$$
$$\sigma P7 = 0$$

Interactions in a Corporate Hierarchy

These are the shared activities between units, where one is not the boss of the other. These interactions are designated as dotted lines in the hierarchy.

For example, in the simple structure shown here,

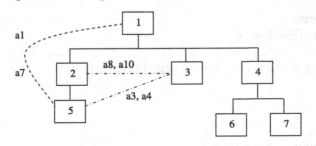

Interactions:

The More Specific (Sophisticated) Calculations Involving Methods I and II

1. *Construct the Corporate Matrix*

(1) Interview each major unit in the organization to obtain the following information:

- Organization's name;
- Your unit's name;
- Your unit reports to which unit (your "boss");
- Units which report to your unit;
- Percent of the entire organizational budget which your unit controls;
- Major activities (one line per activity).

- Name and title of the person leading this unit

  Address;
  Phone/Fax/E-mail;

- Name of the person conducting this interview;

  Date;
  Phone/Fax/E-mail;

(b) The result of the interviews is the organizational matrix.

Unit	a1	a2	a3	a4	a5	a6	a7	a8	$\sum ai$
				Major Activities *ai*					
1	5	5	5	5	5	5	5	5	40
2	3	3	0	1	5	5	0	3	20
3	0	5	1	0	5	3	1	5	20
4	1	1	0	0	0	0	3	5	10
5	5	0	0	3	0	1	1	0	10
6	0	0	5	0	0	5	0	0	10
7	0	3	3	3	1	0	0	0	10

$\sum ai$ = the sum of the major activities, *ai*, for a unit.
5, 3, 1, 0 = the value of the major activity, *ai*, for unit *i*, where:
5   is very important,
3   is of average importance,
1   is unimportant,
0   indicates the units do not do that major activity.

2. Designing an Optimal Corporate Structure: Methods I and II

A. Electron shells in an atomic structure

Each shell has a certain total energy content.

The shells are defined by an "*n*" number. Those electrons with the same "*n*" value are in the same shell.

The total energy content for the shells increases for locations further from the nucleus.

Electrons filling the various shells may have different energies. Electrons fill the inner shells first before filling the outer shells.

B. Designing an Optimal Corporate Structure: Methods I and II

Method I

Each shell has the same total activity content $\sum (\sum ai)$, where $\sum ai$ is the activity level for each unit in that shell, and $\sum (\sum ai)$ represents the sum of the activity levels for those units. From the corporate activity matrix, we obtain the activity level, $\sum ai$, for each unit.

The president's unit alone occupies Shell 1, with its $\sum ai$. All subsequent shells need to have their $\sum (\sum ai)$ approximately equal to that of Shell 1.

Starting with those units with the highest $\sum ai$ values, fill Shell 2 until its $\sum (\sum ai)$ is approximately equal to that for Shell 1. Fill Shell 3 similarly, until its $\sum (\sum ai)$ is also approximately equal to that of Shell 1. The corporate structure is now established, as are the number of shells. This allows the calculation of Pi, Ci and fi for each unit in the corporate matrix. Determining Pi, Ci and fi yields values for the six Hershey parameters, and the bar graph showing these values, and establishes the particular Hershey lifecycle for the corporation as a whole, and the Hershey lifecycles for the six individual parameters.

Method II

To belong to a particular shell, a unit needs its activity level, $\sum ai$, to conform to the range of $\sum ai$ for that shell.

First it is necessary to establish the number of shells. This is accomplished by summing the $\sum ai$ for every unit in the corporation, which yields the Total $\sum ai$. Then divide this Total $\sum ai$ by the $\sum ai$ for Shell 1 (the president). This ratio shows the activity expended by the rest of the corporation in support of the activities of the president in Shell 1. The ratio is, by definition, the number of shells in the corporation. A large value for this ratio indicates more shells are needed to do the work of the corporation whereas a small ratio means the opposite.

With the number of shells established, the activity range of $\sum ai$ for each shell can be determined, i.e., the $\sum ai$ required for a unit in order to belong to that shell.

For example, if the Total $\sum ai$ for all units is 217, and the $\sum ai$ for the president in Shell 1 is 50, then $217/50 = 4.3$, which means there should be 4+ shells, or actually, a total of 5 shells. This means the $\sum ai$ activity ranges for this structure, by Method II, should be given below.

Shell	$\sum ai$
1(president)	41–50
2	31–40
3	21–30
4	11–20
5	1–10

Now, as in Method I, with each unit placed in its appropriate shell, Pi, Ci and fi can be calculated for each unit in the corporate matrix.

This again yields values for the six Hershey parameters and the bar graphs showing these values, and establishes the particular Hershey lifecycle for the corporation as a whole, and the Hershey lifecycle for the six individual parameters.

C. *Comparing Methods I and II*

Units	$\sum ai$
1(president)	40
2	20
3	20
4	10
5	10
6	10
7	10

Method I

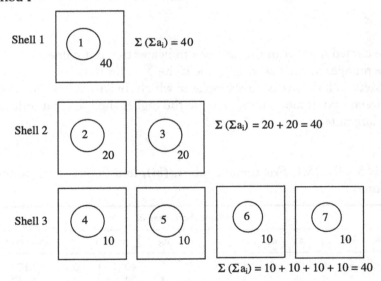

Shell 1   1   40 $\Sigma\,(\Sigma a_i) = 40$

Shell 2   2   20   3   20 $\Sigma\,(\Sigma a_i) = 20 + 20 = 40$

Shell 3   4   10   5   10   6   10   7   10

$\Sigma\,(\Sigma a_i) = 10 + 10 + 10 + 10 = 40$

Method II

Total $\sum ai = 40 + 20 + 20 + 10 + 10 + 10 + 10 = 120$
Shell 1, $\sum ai = 40$
Ratio $= 120/40 = 3$

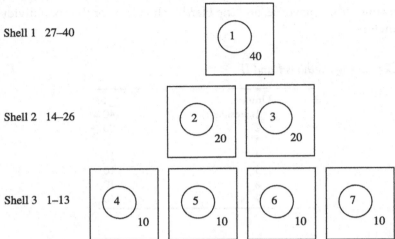

Σaᵢ Range

Shell 1   27–40

Shell 2   14–26

Shell 3   1–13

Note:

- The circled number in the unit box indicates the unit name.
- The number in the lower right side is the $\sum ai$ for that unit.
- In Method I, if there is a question as to which shell to assign a unit (a tie between two or more units), choose the higher shell for that unit with the largest bi.

3. Add Shells, (Si), Fractional Budget, (bi), and Power, (Pi), to the Matrix

Unit	a_1	a_2	a_3	a_4	a_5	a_6	a_7	a_8	$\sum ai$	Si	bi	$Pi = bi/Si$
1	5	5	5	5	5	5	5	5	40	1	0.1	0.100
2	3	3	0	1	5	5	0	3	20	2	0.1	0.050
3	0	5	1	0	5	3	1	5	20	2	0.2	0.100
4	1	1	0	0	0	0	3	5	10	3	0.2	0.067
5	5	0	0	3	0	1	1	0	10	3	0.2	0.067
6	0	0	5	0	0	5	0	0	10	3	0.1	0.033
7	0	3	3	3	1	0	0	0	10	3	0.1	0.033

4. Boss relationship (for both Methods I and II)

The corporate structure looks like the following, with the shared activities shown. The numbers associated with the activities are the weighting factors (5, 3, 1, and 0) for that unit for which a boss is to be determined.

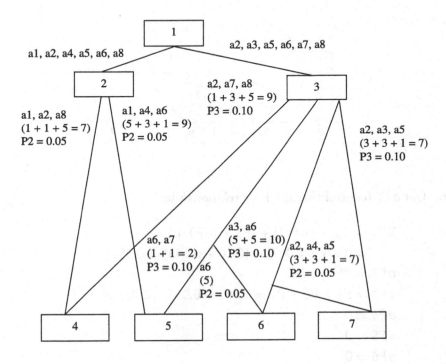

- The boss for Units 2 and 3 is Unit 1
- The boss for Unit 4 is Unit 3
- The boss for Unit 5 is Unit 2
- The boss for Unit 6 is Unit 3
- The boss for Unit 7 is Unit 3

Note: To find the boss, get common $\sum a_i$ for possible bosses. Multiply this number by P_i for the possible boss. Choose as a boss the candidate with the largest $\sum a_i \times P_i$ value.

5. Redraw the corporate structure, using solid lines for boss relationships and dotted lines for interactions (shared, non-boss activities)

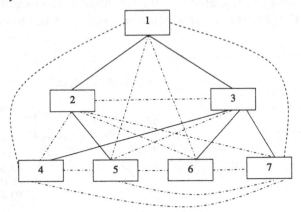

6. Get σPi (control through boss relationships)

$$\Sigma P1 = [P2 + (\sigma P2)] + [P3 + (\sigma P3)]$$
$$= [0.05 + 0.067] + [0.10 + 0.133] = 0.35$$
$$\sigma P2 = P5 = 0.067$$
$$\sigma P3 = P4 + P6 + P7 = 0.067 + 0.033 + 0.033 = 0.133$$
$$\sigma P4 = 0$$
$$\sigma P5 = 0$$
$$\sigma P6 = 0$$
$$\sigma P7 = 0$$

7. Complete the matrix, adding $\sigma P_i, C_i$ and f_i columns

Unit	a_1	a_2	a_3	a_4	a_5	a_6	a_7	a_8	Σai	Si	bi	Pi	σPi	$C_i =$ $(Pi + \Sigma Pi)$	$f_i =$ $(Ci \Sigma Ci)$
1	5	5	5	5	5	5	5	5	40	1	0.1	0.10	0.350	0.450	0.450
2	3	3	0	1	5	5	0	3	20	2	0.1	0.05	0.067	0.117	0.117
3	0	5	1	0	5	3	1	5	20	2	0.2	0.10	0.130	0.230	0.230
4	1	1	0	0	0	0	3	5	10	3	0.2	0.067	0	0.067	0.067
5	5	0	0	3	0	1	1	0	10	3	0.2	0.067	0	0.067	0.067
6	0	0	5	0	0	5	0	0	10	3	0.1	0.033	0	0.033	0.033
7	0	3	3	3	1	0	0	0	10	3	0.1	0.033	0	0.033	0.033

$$\Sigma Ci = 0.997$$

8. Interactions (dotted lines between units)

Unit–Unit
1-4, 1-5, 1-6, 1-7
2-3, 2-4, 2-6, 2-7
3-5
4-5, 4-7
5-6, 5-7
6-7

9. The six Hershey parameters, bar graph, lifecycle status and lifecycle status of the organization as a whole

Parameters	Numerical Values	Lifecycle Status
SR-A	0.800	Old-Bureaucratic
SR-G	0.467	Prime
SR-I	0.305	Prime
PD-CG	0.611	Mature-Stable
PD-SN	0.838	Old-Bureaucratic
ST/ST$_{max}$	0.664	Mature-Stable
The organization as a whole (the sum of the six parameters)	3.685	Mature-Stable

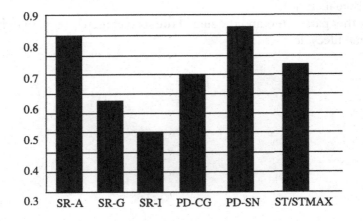

Case History
University of Cincinnati
Nov. 15, 2005
A Study of the Organizational Structure of the College
    Conservatory of Music (CCM)

By
Daniel Hershey (Ph.D. Professor of Chemical Engineering)
Munish Gupta
Vijay Wagh
Raad Asaf

1. Perception
2. Method II

Summary of the Study (Refer to Pages 211–215)

Perception
- There are four shells (levels).
- All units report directly to the Dean.
- The Assistant Deansn in Shell 3 seems under utilized.
- There seems to be a significant degree of overlapping activities (SR-A, Old-Bureaucratic).
- The other parameters and the sum of the six parameters seem to indicate a Prime lifecycle.

Method II: Theoretical Model

- There are seven shells (levels).
- Shells 2, 3 and 7 are empty. Some units need to be given more responsibilities to qualify for Shells 2 and 3.
- Most units report directly to the Dean.
- It seems that the theoretical boss for Unit 4 (Assistant Dean, Admission and Student Services) and Unit 6 (Director, Preparatory Department) is Unit 12 (Head, OMDA Division).
- It seems that the theoretical boss for Unit 3 (Assistant Dean, Administration and Performance Management) and Unit 5 (Director, Development and External Relations) is Unit 4 (Assistant Dean, Admission and Student Services).
- There seems to be a significant degree of overlapping activities (SR-A, Old-Bureaucratic).
- There seems to be a high degree of verticality in the table of organization (SR-G, Old-Bureaucratic).
- The other parameters and the sum of the six parameters seem to indicate a Mature-Stable/Prime Lifecycle.

Summary of Results

1. Perception
   (What they think is true)

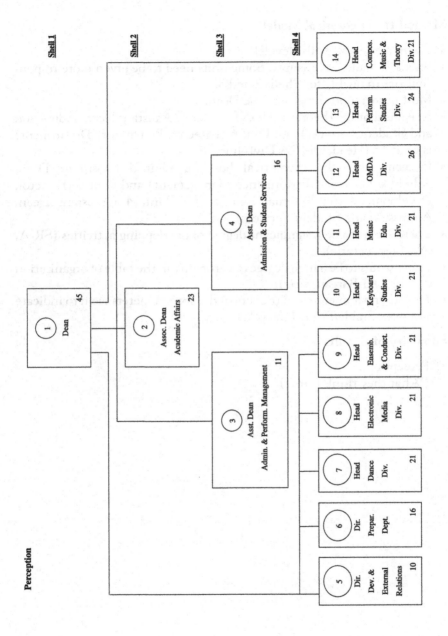

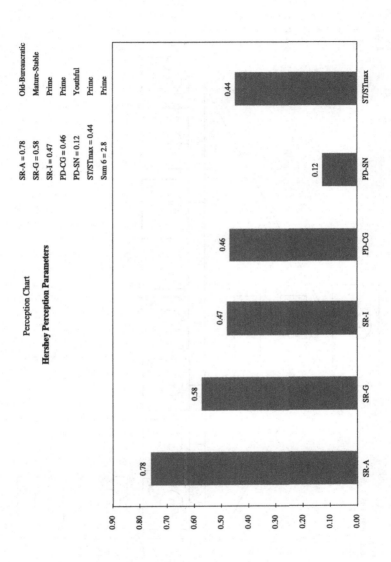

Perception Chart

Hershey Perception Parameters

SR-A = 0.78	Old-Bureaucratic
SR-G = 0.58	Mature-Stable
SR-I = 0.47	Prime
PD-CG = 0.46	Prime
PD-SN = 0.12	Youthful
ST/STmax = 0.44	Prime
Sum 6 = 2.8	Prime

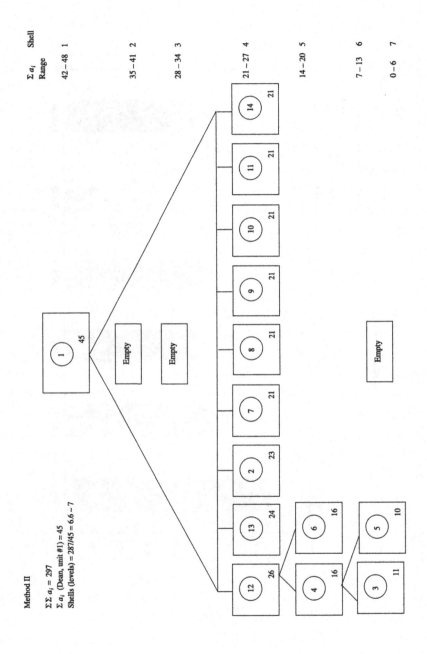

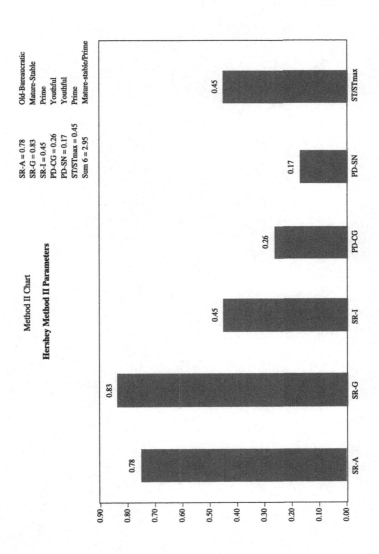

Method II Chart

Hershey Method II Parameters

SR-A = 0.78	Old-Bureaucratic
SR-G = 0.83	Mature-Stable
SR-I = 0.45	Prime
PD-CG = 0.26	Youthful
PD-SN = 0.17	Youthful
ST/STmax = 0.45	Prime
Sum 6 = 2.95	Mature-stable/Prime

PART VI

Entropy Theory at Aging Systems: The Universe

CHAPTER 16

The Universe and Beyond

An Expanding Universe?

Opinions vary as to whether the expanding universe will continue to enlarge forever, or that it will expand to some sort of critical size and critical density, and remain at that size (the steady state universe). Or that, after the expansion, it will contract to some miniscule remnant (the big crunch), diminished by gravitational forces. And perhaps after the big crunch, the universe will experience another big bang and repeat the cycle again, and again, and again.

Seeing the universe as a cycling entity suggests some sort of analogy with the game played with a wooden paddle and a rubber ball tethered to the paddle by a rubber string. Hit the ball with the paddle and it flies out towards some outer limit (which depends on the initial velocity of the ball and the elasticity of the string, among other factors). The ball may then be pulled back to the paddle by the contraction force exerted by the rubber string (like gravity), or if conditions are right, the string breaks and the ball may continue its flight away from the paddle.

If the universe were to expand forever, the unasked and hence unanswered question is, where does the universe go if it expands forever? Will it disappear eventually? Into what? Does the universe have a confining envelope around it?

If the universe were to achieve some critical size and remain that way, then it would seem to have a boundary and we can ask, what is beyond the boundary? If after expanding, the universe were then to begin contracting, would it become an incredibly small, dense fluid, gas, or solid? And disappear?

If the present universe (considered as a gaseous sphere), contracted to a miniscule, dense nuclear fluid, it might end up with a density of

10^{14} gm/cm^3, while by comparison, it is estimated that the present density of the universe is 10^{-30} gm/cm^3. Some say this dense nuclear fluid might be neutrons and electrons compressed to protons. Estimates for the radius of the universe put the figure at 20 billion light years.

An Empty Universe?

Some say the empty space of the universe is filled with "ether", a gaseous mass which is called invisible dark matter. They say this dark matter exerts a gravitational pull which will eventually cause the universe to end its expansion and begin its contraction.

Some say the universe is a bubble in infinite space, a singularity, an open system of mass and energy, and information, capable of exchanging mass, energy, and information with other universes. Or with whatever ambience surrounds the universe. Or is the universe a closed system where nothing diffuses through its borders?

A Hot and Cold Universe?

The temperature of the universe is determined by cosmic infrared microwave radiation. The velocities of its internal bodies are inferred from their red shift. Upon birth, the universe is believed to have been very hot, and with age, and expansion, has cooled to where the temperature of the interstitial space of the universe is very near zero degrees Kelvin (absolute zero). If the universe continues to expand and cool, it may reach absolute zero temperature. And will this be the end of the universe? Heat death? Will it disappear?

A Big Universe?

It is estimated that the present age of the universe is between 8 and 20 billion years, and it contains clusters of stars and other material making up galaxies, of which there may be 100 billion. The average distance between galaxies may be 2 million light years. Our galaxy is considered a circular disc, 80,000 light years in diameter and 800 light years deep. Others suggest a diameter of 100,000 light years and a depth of 400 light years. The universe may now be expanding 5 to 10 percent every billion years.

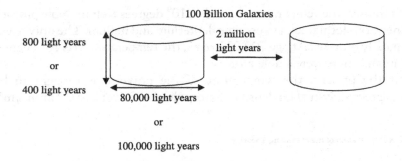

800 light years

or

400 light years

100 Billion Galaxies

2 million
light years

80,000 light years

or

100,000 light years

The Temperature, Chemistry, and Physics of the Universe.

It is believed that hydrogen constitutes 93 percent of the total atoms in the universe, or about 76 percent of the total mass. Helium is present in lesser amounts, constituting 7 percent of the total atoms, or about 23 percent of the total mass. In the beginning, about one-half hour after the big bang, hydrogen and helium constituted 99 percent of the total matter; deuterium amounted to about 1 percent. Theories hold that the universe began from a core of neutrons, which under expansion decayed to protons. Then the protons captured neutrons, forming deuterium. Deuterium captured another neutron and became tritium or another form of hydrogen, which eventually transmuted to helium.

In the beginning, the universe was at a very high temperature and pressure. As the universe expanded and cooled, its dense matter is believed to have become gas clouds. Some believe in this cooling universe, there was a transferring of energy to the gravitational fields.

The universe temperature may have been 250 million degrees Kelvin after one hour of existence, 6,000 degrees Kelvin after 200,000 years and 100 degrees below the freezing point of water at age 250 million years. About 300,000 years after the big bang, the universe was a dense, hot haze of subatomic particles that hadn't yet coalesced into atoms. Later, stars and galaxies were born. Astronomers refer to the period before stars and galaxies were born as the dark ages.

The big bang of the universe is like a bubble of steam forming in a pool of heated water. After 0.01 seconds, the temperature of the universe may have been 10^{11} degrees Kelvin. The universe consisted of matter and radiation (photons, electrons, positrons, neutrons). At one second, the temperature was 10^{10} degrees Kelvin and the universe contained 10^{65} grams of mass.

At 3 minutes, the temperature cooled to 10^9 degrees Kelvin. Now protons and neutrons form nuclei to become deuterium and helium. The universe is still mostly protons. At age 700,000 years, the temperature was 10^3 degrees Kelvin and there were stable atoms.

At the present, the temperature of the universe is thought to be 2.73 degrees Kelvin. The density of atoms in the universe is 0.2 atoms/m^3.

An Illustration of the Prevailing Theories

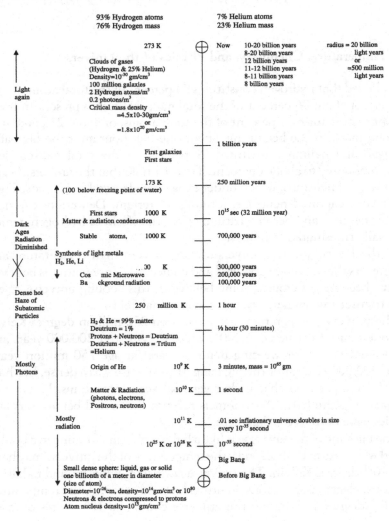

A Summary of the Prevailing Theories

Hubble believed in 1929 that the universe was expanding, and that the speed of expansion was declining and that empty space was not empty, but is filled with radiation and mass.

Some believe the density of the universe ranges from 1.8×10^{-29} gm/cm^3 to 4.5×10^{-30} gm/cm^3. If above this density, the expanding universe may begin to contract, returning to the very dense, spherical core from whence the universe sprang during the big bang. This dense, spherical core could be 10^{-26} cm in diameter, with a density of 10^{80} gm/cm^3, compared to the density of an atomic nucleus of 10^{15} gm/cm^3.

Others believe the big bang, sometimes called the inflationary universe, caused an expansion against essentially zero pressure and temperature.

Or the universe may have been born out of a vacuum or black hole, and may have no fixed boundaries. It may have been formed from photons, which were background radiation, and then became mass. In this view the universe is then a self-organized, open system, in a non-equilibrium state. And life represents an improbable and transient fluctuation.

CHAPTER 17

Infinity, Entropy and the Aging Universe

Infinity and the Universe

I'll count my numbers. Count with me: 1, 2, 3, 4 . . . Give me your largest number. Is it one billion? One zillion? Or if you work with numbers and play mathematical games, is it $10^1, 10^2, 10^3, 10^4, \ldots$? Stretch for the largest number you can imagine, and I can add one to it. One zillion you say, and I respond with one zillion and one. One zillion zillion, you say, and I counter with one zillion zillion and one. But where is Infinity? Find it for me. Point to it. Touch it! Infinity is big, and bigger and biggest. It is large, and larger, and largest.

And who is out there in Infinity? Is there someone or something out there in Infinity? Who lives there, we would like to know. If you lived in Infinity, you would have to be very intelligent to have found your way there, to live there, to understand how to exist in Infinity. Like the spy who came in out of cold, it must be possible to be transported out of Infinity, to a smaller place, which for us means our not-Infinity existence, our finite universe. In Infinity, it must take great power, almost infinite knowledge, to reside there.

Here is an infinite series of numbers: 1, 2, 3, 4 . . . We can sum this series by adding the second term (2) to the first term (1) to get 3. Doing it again, we can add the third term (3) to the sum of the first to terms (3) to get 6. Then once again, add the fourth term (4) to the sum of the first three terms (6) to yield 10. Summing the original infinite series 1, 2, 3, 4, . . . produces another infinite series 1, 3, 6, 10, . . . In other words, the sum of the original infinite series is itself an infinite series, which says the sum has no limit: adding the next number to the sum of the previous numbers always produces a larger number than before, as we write more terms. There is disorder out there in Infinity, if the sum of an infinite series of numbers is itself infinite.

On the other hand, another infinite series of numbers, $(1/2)^1, (1/2)^2,$ $(1/2)^3 \ldots (1/2)^n$ can also be summed out towards Infinity, and we can show that the sum of these terms is equal to 1.00. So, in some cases, going out to Infinity yields maximum disorder, while in other cases, traveling out to Infinity yields something finite, an ordered situation. We say this series, $(1/2)^n$, converges, and the sum of the Infinity of terms is equal to 1. Thus we have figured out how something behaves in the vicinity of Infinity. And we found structure and information. We are able to determine the sum; the sum converges in the vicinity of Infinity.

This series, $1, (1/2), (1/3) \ldots (1/n)$ behaves in an opposite way, for the sum of an Infinity of these terms isn't a finite number at all, and we say the sum goes to Infinity. We say this series of numbers diverges. We know how this particular something behaves in the vicinity of Infinity. We find it doesn't converge; it diverges, which means there is no structure and no useful information.

We may conclude from these $(1/2)^n$ and $(1/n)$ examples that systems can behave differently in the vicinity of Infinity. Some things are clear there; some things are not. We can see that $5/x$ goes to zero when x gets very large: very, very large: infinitely large. It is also clear that $x/5$, the inverted form of $5/x$, behaves very differently. When $x = 10$, $x/5 = 2$, when $x = 100$, $x/5 = 20$, when $x = 1000$, $x/5 = 200$. But what is value of $x/5$ when x gets very, very large, approaching an infinite number. When this happens $x/5$ approaches an infinite number. We say it approaches Infinity. Dividing Infinity by 5 still gives Infinity. But what happens in this case, $x/2^x$, when x gets very large, approaching Infinity. Both numerator and denominator will get very large, each approaches Infinity. Yet, there is something different at work in this example. We might sense that the numerator, x, will tend to be smaller than the denominator, 2^x. And even when we let x be a very large number, approaching an infinite value, this still will hold true. We can show mathematically that the 2^x Infinity will dominate the x Infinity, so $x/2^x$ will go to zero when x goes to Infinity. The lesson here is that apparently we can have two different magnitudes of Infinity.

What can we derive from, $2^x - x$, when x goes to Infinity? We have apparently Infinity minus Infinity. We are subtracting an infinite number from another infinite number. Is the result zero, Infinity, or something in between? Again we can show that the two infinites are different, that the 2^x Infinity will dominate the x Infinity and the result is $2^x - x =$ Infinity when x goes to Infinity. Again the lesson to be learned is that not all Infinities are alike, not all Infinites are the same size, and that there are lesser Infinities which yield to the greater Infinity. That there is a super Infinity, within which all lesser Infinities are subsumed.

Infinity is nothingness, and yet it is everything. It has no dimensions, no corporeal essence, and no smells. It is the stuff of energy and information. It encompasses all voids. It has no limits, no end, no beginning. It just is. Infinity doesn't depend on the human brain for visualization. It is beyond the human brain's comprehension. It is a "heavy" religious idea, synonymous with God-like characteristics. But it is out there, somewhere, because we know this:

$$1, 2, 3, 4 \ldots \rightarrow \text{Infinity}$$

$$\text{Birth, Growth, Maturity, Aging, Death} \rightarrow \text{Infinity}$$

Infinity contains an infinite amount of potential information.

Infinity contains all the information we know, or can ever know, or can never know.

Let us once again examine an ordinary book, with its letters organized into words, which are strung in sentences on the pages, which are bound together in a book, whose volume is contained in a library of books. I can rip these pages from the book, and with my scissors cut each page into individual words and each word into individual letters. I can pile these individual letters on the floor. And what have I? I have much more than just a pile of junk letters, the remnants of the organized assembly of letters, which became the words and the sentences for the plot of the story in the book. For what that pile of individual letters contains is a near Infinity of potential information, the potential for writing not only the original book, but many, many other books. Simply by rearranging the letters piled on the floor. That pile of letters is like the infinite potential information of Infinity, from which is obtained the organized and stored information of our book. The book and the library represent the finite world of our limited intelligence.

What does Infinity look like? Is it a Moebius surface?

Can you describe Infinity?

Can you describe maximum disorder?

Is there a physical reality associated with Infinity?

What does death look like? Like Infinity?

Is there something bigger than Infinity?

Is one Infinity larger than another Infinity?

Is there just one Infinity?

What would we see if we got large and larger and larger, approaching infinite size?

What would we see if we got smaller and smaller and smaller, approaching an infinitesimal size?

We have great difficulty "seeing" Infinity. Mostly it has to do with the limitation of our human brain. Mostly we exist through our finite senses of touch, sight, smell, taste and hearing. Something exists for us if it has dimensions, heft and significance. Even an intangible concept, or an emotion, can be defined in terms of its impact on our senses. Our thoughts are circumscribed by our experiences, allowing us to corral these esoterica within a fence, or wall, or membrane. We need to do this in order to understand — or try to understand — our existence. We need to make sense of life and death.

Infinity has more to do with a sense of "being", and that is hard to deal with. Infinity just "is".

To understand Infinity is to deny boundaries.
To understand Infinity is to deny ends and endings.
To understand Infinity is to deny beginning and beginnings.
To understand Infinity is to deny place and places.
To understand Infinity is to deny focus and focusing.
To understand Infinity is to deny existence and existing.
To understand Infinity is to deny limits and limitations.
To understand Infinity is to allow softness and vagueness.
To understand Infinity is to allow amorphousness and amorphosity.
To understand Infinity is to allow expansions beyond limits.
To understand Infinity is to allow that all concepts of God are valid.
Infinity surrounds.
Infinity envelops.
Infinity fills.
Infinity contains.
Infinity controls.
Infinity is beyond.
Infinity is around.
Infinity is inside.
Infinity is up and down.
Infinity is inside and outside.
Infinity is hot and cold.
Infinity is wet and dry.
Infinity is rough and smooth.

Infinity is here and there.
To understand Infinity, don't try to imagine what it looks like.
To understand Infinity, don't try to imagine where it is.
To understand Infinity, begin with an ether which fills all voids.
To understand Infinity, focus on a sense of being.
Infinity must be a universal something, and a universal someplace, and a universal essence.
Infinity is every thing we know, or can ever know, or can never know.
Infinity is here, there, everywhere.
Infinity is wherever we dream, wherever we can dream, wherever we can never dream.
Infinity is hot and cold, hotter and colder, hottest and coldest, upper and lower, uppest and lowest, beyond what we can never imagine.
Infinity is the colors of the rainbow, and colors beyond our ability to see them, beyond what we can never see.

Entropy and the Universe

To understand Infinity is to allow that the laws of entropy apply.

To understand Infinity is to allow that maximum entropy controls ingress and egress.

Infinity's porthole or doorstep is maximum entropy. We enter Infinity through maximum entropy.

We elide through maximum entropy into Infinity.

Entropy can tell us something about time's arrow, about the direction of time. Measure my entropy now and ten years later. I will look older in ten years and my entropy will be higher than it is now. Entropy maps the degree of order and disorder: higher entropy indicates more disorder, which means I can calculate my entropy ten years hence and confirm that its higher value shows time has passed, from my present to my older future. I can tell which came first, the green leaf of spring or the red of fall, by calculating the green leaf entropy and the red leaf entropy. The red will be greater, telling me from the second law of thermodynamics that this leaf has evolved from lower to higher entropy, from order to disorder, from spring to fall. I know spring came first because spring green has less entropy than fall red.

We are born, live for a few years, and die. We are born, grow through the juvenile phase, stop growing, and mature, evolve through adulthood, and become old, older, aged and die. We are born organized, age with increasing disorder, and die in maximum disorder, unable to control or reduce the disorder. We are born with low entropy, mature with increasing entropy, and die with maximum entropy. We are born far from maximum entropy, and by aging, we get closer to maximum entropy, and die in the vicinity of maximum entropy. The game of life, then, is to keep our distance from maximum entropy. The driving force for life is the entropy distance from maximum entropy. The requirement for long life is to eat and work properly, to keep a healthy gulf between us and maximum entropy, as deep as possible, for as long as possible.

Death is the ultimate disorder, maximum entropy. Have you recently walked behind a very old person and observed how stiffly they amble? It is a body out of control. A simple cold, so easily overcome by the young, becomes a killing stress in the old. Athletes cannot maintain their skills forever. Sooner or later, reaction time erodes, strength diminishes and energy subsides. Entropy increases. It is all about entropy.

Achieving maximum entropy is a randomizing process. It means landing in the most probable state, the universal attractor, with the most probable distribution of energy, atoms, molecules, beings, things, worlds, universes. It means having our useful stored information converted to less useful potential information.

Entropy characterizes symmetry. More symmetry, higher entropy. More randomization.

Entropy characterizes concepts. An expanding universe proceeds with increasing entropy.

The origins of life, the evolution of life, began as a self-organizing lurch to low entropy. Setting off a ticking clock of life and existence. When we achieve maximum entropy, it ends.

Life expectancies and the finiteness of lifespan and the factors affecting it are entropy derived.

The orderly decay, the aging process, is entropy driven, towards maximum entropy.

Self-organizing communications networks are entropy driven.

Information flows in biological systems which organize themselves are entropy driven.

Dissipative structures are entropy driven.

Maintaining life far from equilibrium is entropy driven.

Choosing a life path at a bifurcation point is entropy driven.

The success of corporate structures is entropy driven.

The decline and fall of civilizations is entropy driven.

The nature of civilizations, the birth of civilizations, and the death of civilizations are entropy driven.

A gain in entropy means a loss of information.

Entropy describes disorder and our ignorance.

Entropy tells us the direction things will go.

Entropy says we cannot take heat and convert it all to work.

Entropy says that when we mix things, there results more disorder, loss in stored information, more randomness.

Maximum entropy is when everything is equal, when there is total randomness.

It would seem that the entropy laws here on earth dictate that we have a certain lifetime potential of entropy, and we live our lives spending this capital. This means we accumulate entropy faster or more slowly, depending on who we are and what we do. But sooner or later we accumulate that critical maximum entropy which defines death. Our actual age, then, is not necessarily our chronological years, but is our entropy age. Our entropy age represents our entropy accumulation, while Excess Entropy tells us about our entropy distance from maximum entropy. Excess Entropy Productions provides information on the speed at which we are living and approaching maximum entropy.

It would appear that our universe is programmed, limited by maximum entropy, beyond which we cannot exist. We are of an age defined by Excess Entropy, and characterized by Excess Entropy Production. We expect our universe to end when we achieve maximum entropy. Our task then, is to calculate the entropy content of our universe now, and to calculate the limiting maximum entropy for our universe. This yields the lifespan potential for our universe, and we can track over time Excess Entropy and Excess Entropy Production for our universe. We will know our universe's entropy age, and the nature of our dissipative life (the velocity at which we are accumulating entropy).

All of this we determine from the second law of thermodynamics, with the acknowledgement that all of this is subject to change if there is outside intervention in the affairs of our universe, from forces beyond our universe. Perhaps these interventions will allow our universe to evolve, Darwinian style, into a new universe, further from maximum entropy than before. Perhaps the nature of our universe, perhaps our basic laws, would then change, yielding new, lower entropy content in this new stationary state, and more time to live.

Entropy is the measure of disorder in a system.

Entropy increases as differences or tensions within the system are dissipated.

Entropy increases as size increases.

Entropy tends towards a maximum in the vicinity of death, as control is lost.

We tend towards maximum entropy when all differences and tensions disappear (a disaster or chaos condition).

Excess Entropy (EE) is measure of the entropy distance from disaster. EE approaches zero as we approach disaster or chaos or death.

A system in this state of disaster is at maximum entropy. We can calculate $(S - S_{max})$, the entropic distance from disaster or death, which is Excess Entropy (EE). We can also calculate the speed of approach to disaster, Excess Entropy Production (EEP). Evolution (like Darwin's theory of the survival of the fittest) yields more viable forms of life with lower entropy and greater EE values. Mutations in nature, in living systems, may be a way of forcing a greater distance between entropy and maximum entropy, and giving new life (rejuvenation) to a species.

Closed systems do not self-organize, they self-destruct. This means we need to ensure that all the things we do, all the systems we work with, are open. We need to remain receptive to the exchange of information, energy, and mass between the external environment and the internal working units. We need to pay attention to the size of our organization, the geometry of the internal structure, the duration and intensity of the forces for change, the degree of exchange of information within the organization, the nature of the external environment, and the quality of the information available. Is the border or membrane surrounding our organization stable and porous?

Self-organizing systems evolve through spontaneous increases in organizational control and complexity. A property of self-organizing systems is the ability to learn, that is, self-modification, which is a spontaneous emergence of order, far from equilibrium. The convective flow pattern, Bernard cells, formed by water being heated in a shallow pan, is an example of a self-organizing system. The totality of life on earth behaves in some ways like a living organism, and seems to regulate itself in the short run (constant oxygen content in the atmosphere or the salinity of the oceans). But we know that the present self-organizing earth state is simply one of an infinite stationary or steady states possible. It is only a matter of waiting long enough to see the evolution from one stationary state to the next. For example, there can be biological evolution, as well as self-organizing and evolving linguistics, crystal growth, social processes, committees, and task forces: all self-organizing and evolving.

If a salt solution of sodium and chloride ions in water dries out, what remains are salt crystals, sodium and chloride ions now locked in an unalterable structure. Self-organization. Might life have begun so simply, where from some sort of primordial soup there arose some combination of matter which self-organized into something functioning, eventually something capable of reproduction? Self-organization and then evolution into other forms. The formation of structure is a key to self-organization.

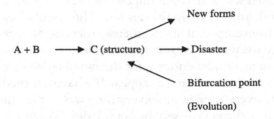

But what drives the changes? What forces us to change? Is it a social force, or the power of numbers present, or is it an entropy force, directing us to pay attention to order and disorder, and to evolve to new forms far from maximum entropy?

We are dissipative structures, all of us: people; water; buildings, etc., which means we are continually generating entropy. We rest in a stationary state which represents an efficient, lowest level of entropy production, waiting for the next entropy push, to another place, another structure. And we do it in a self-organizing way, driven knowingly, towards the next stationary state, to rest again until driven again. Evolution–Revolution, the driving force for change, designed to keep us beyond maximum entropy's reach a bit longer, designed to allow us to enter another entropy stationary state. Until ...? Until, inevitably, we find our way to maximum entropy and death.

We the living are born as dissipative systems, entropy generating, growing, differentiating, and changing. We are being driven to change, self-organizing, by forces hormonal. We are programmed by genes conditioned through the ages, to be formed a certain way. To evolve. Ontogeny recapitulates phylogeny. The forces drive us far from equilibrium so we cannot return to where we were. We are driven beyond the point of return, beyond the status quo. Far from equilibrium, at this bifurcation point, most of us find the path towards the next stage of our development, from childhood to young adulthood and maturity. The other bifurcation path leads to disaster, ill functioning bodies and perhaps early death. At each stage in our development, at each stationary state, we hunker down, collect ourselves, come to grips with what we have become, "smooth our feathers", minimize our entropy production, and get into a steady, comfortable state. We are open systems awaiting the next push towards the next level of development. We mature, age, become aged, and head relentlessly towards maximum entropy, as we change in appearance and change in efficiency of internal body operations. Eventually we accumulate the requisite maximum entropy, signaling a degree of disorder which makes it impossible to control the life forces (the chemistry of life), and maintain the tension of life. And so we die. Death is the transition, from corporeal life to ethereal essences. Now the membrane, the boundary, to the infinite becomes permeable. It is a necessary condition, this condition of maximum entropy, for the tangible to cease functioning, for the essences to be released and disappear. We have returned from whence we came, where everything is, and everything was. At maximum entropy, everything returns, through dissolving boundaries. We exist as singularities, discontinuities, temporary escapees. To be returned somewhere when conditions are right. Driven by entropy, Excess Entropy, and Excess Entropy Production.

The big bang theory for the birth of our universe hypothesizes an expanding existence, into the infinite. All accomplished in infinite time. On earth, as time evolves, things may settle down into a steady state, or into a fixed condition. Or, things may not settle down, may nor converge at all, but may diverge and reach maximum entropy and drift away. Time is governed by entropy and Excess Entropy and Excess Entropy Production and maximum entropy. On earth, liquid water freezing to solid ice, or gaseous steam condensing to liquid water, change entropy content significantly. Ice (low entropy), water (higher entropy), steam (highest entropy) trace the transition and evolution and phase changes from order, to disorder, to maximum disorder. We can calculate the entropy increase, from ice, to water, to steam. And where will the steam go as its entropy continuous to increase, as we heat it more and more? Into Infinity?

There may be parallel histories, between us, as living human systems, and the universe, also a living system: Our human birth (or the big bang of the universe), followed by growth, aging and death. Then in death at maximum entropy, maximum randomness and disorder, loss of stored information, we (and the universe) are returned from whence we were derived (the earth and sky for us, Infinity for the universe).

Infinity, Entropy, and the Universe

If Infinity is entered through the porthole of maximum entropy, and total randomness and the transformation from total stored information to total potential information, then how did we reverse this process and become our universe? How did we separate from Infinity and its infinite potential information?

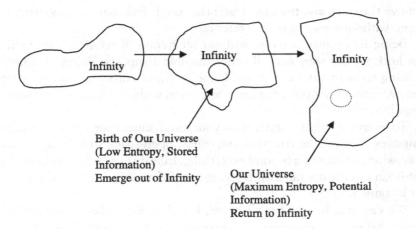

Scientist believe our world, our universe, began with a big bang, but the bigger question is, where did the big bang come from? Within Infinity (which contains infinite potential information at high entropy) it is possible for pockets of low entropy to form spontaneously, to self organize. It was one of these vacuoles or islands of low entropy which became our universe, our world, our galaxy, our Milky Way. And isn't it logical that there would be other universes in existence, presently beyond our knowledge or comprehension. We are ephemeral states, existing so long as your content stays below maximum entropy. Inevitably, we achieve maximum entropy and total potential information and "see" the entry port, back to Infinity (as Mary did in the Secret Garden). We dissolve into the infinite potential information ether of Infinity. Within the context of Infinity and its eternity, our universe and our existence on earth are transitory and will be relatively short-lived. If we could calculate what constitutes maximum entropy within the universe, and if we could calculate what is presently its entropy content, this could be very meaningful. The difference between maximum entropy (where we are heading) and entropy (where we are presently) is the distance, in entropy terms, to the end of our world, when we will return to Infinity. This entropic distance is called Excess Entropy. The speed at which we are approaching maximum entropy is called Excess Entropy Production.

Before we existed, there was Infinity. Infinity was everywhere. Infinity will be everywhere. Sit quietly and contemplate Infinity. Picture in your mind an expanse, up to the horizon. Now, in your mind, remove the horizon. It feels like nothingness without the horizon, without some frame of reference. Picture in your mind the sky, up to the stars. Now, in your mind, remove the stars, and the moon, and the sun. It feels like nothing without them, without some frame of reference.

Being in Infinity is to be without all frames of reference, including the horizon, the stars and all other tangible signposts. Being in Infinity is being without limits. It is just being, without corporeal frames of reference, without temporal frames of reference, without physical and mental boundaries.

To be in the infinite state, clear your mind, clear your mind's eye of all tangibles. You will be left with emptiness, open space stretching beyond everywhere, expanding beyond everything. Infinity may at first seem empty, but it isn't, for it contains all essences and information imagined and not yet imagined.

We can ask, how did it all begin, but then we realize Infinity didn't begin. Asking the question demonstrates the limitations of our human brains, which require a beginning, middle and an end.

It is useless to ask, what is it, because Infinity isn't related to anything we can know. Infinity simply is.

It is useless to ask, what is beyond Infinity, because there is nothing beyond Infinity.

It is useless to ask, what are coordinates of Infinity, because Infinity is beyond coordinates.

It is useless to ask, who is in Infinity, because the answer is yes.

It is in Infinity that entropy controls. Potential information is Infinity's currency, its energy, its raw material.

In Infinity, a random accumulation of stored information at low entropy allows the condensation of vacuoles of matter and energy. And in forming, the vacuoles temporarily separate themselves from the substrate of Infinity. It was such a precipitated mass of energy that became our universe, began our lifecycle with a "big bang". Our universe, originating as a stationary state of low entropy and high stored information, has begun to expand and age, generating its own internal ambience and lifecycles. We were born in a low entropy stationary state and have evolved through succeeding stationary states with ever increasing entropy and expansion. Towards maximum entropy, and total potential information. As Excess Entropy and Excess Entropy Production wind down towards zero. At which time the curtain will part and we and our universe of now total potential information will return to Infinity, to dissolve into its infinite milieu of infinite potential information. In the meantime, before this denouement, we ponder the immense questions of God, whose God, how many Gods, and the meaning of life and death, heaven and hell. If the God or Gods of our universe exist, then they too must have evolved from Infinity. Perhaps our Gods were born when our universe was formed. Or perhaps Infinity and God are synonymous.

The laws in Infinity are the laws of information and entropy. The look of Infinity is the look of potential information and entropy.

The universe was born from Infinity when a packet of potential information was spontaneously transformed into stored information at minimum entropy. And we coalesced.

The universe condensed from Infinity much as steam molecules, at high entropy, condense to lower entropy water droplets, and then to the still lower entropy content of ice crystals.

Infinity is filled with ether-like infinite potential information.

In the study of chemical reactions we talk about "activation energy", that is, the degree to which atoms need to be energized or aroused, before they can enter into a new state. In a sense, the atoms need to achieve a certain degree of pressure for change, before the curtain parts or the barriers disappear and the atoms are free to enter into their new, far-from-equilibrium states.

Entropy, being a measure of the disorder contained within us, plays an analogous role. Instead of activation energy in chemical reactions, we speak of maximum entropy as the activation state which allows the curtain to part in Infinity. To depart Infinity and its infinite potential information, our universe spontaneously coalesced and precipitated from Infinity, transformed into a pocket of stored information and low entropy. Thus our universe was born, grows, matures and ages. Upon accumulating maximum entropy, and zero stored information, in old age, our universe will die and return to Infinity.

Our Aging, Evolving Universe: From Birth (Out of Infinity) to Death (Return to Infinity)

The Hershey Model

Infinity is total potential information, like the pages of a book cut into individual lines, and the lines cut into individual words, and the words cut into individual letters, all piled in a heap on the floor. That pile of individual letters on the floor is total potential information. From this we can theoretically reconstruct the original book of stored information, or other books. There is an almost infinite availability of potential information in the pile of letters on the floor. Infinity is like this, only in Infinity there is an infinite availability of potential information: everything we can know, or never know.

With the pile of individual letters on the floor, suppose a wind arises and stirs this pile of letters. It is theoretically possible that this forcing wind can cause the formation of an organized assemblage of the letters, which may be an intelligible book.

Similarly, in Infinity, there could arise an organizing force which acts like the wind effect on the pile of letters. In Infinity, this self-organizing force might generate, from the Infinity of potential information available, a dense nuclear fluid (somewhat like the formation of a cloud in the sky).

The infinite potential information of Infinity could then have a yielded a dense nuclear fluid of stored information. Potential information is associated with high entropy while stored information is a low entropy state. Infinity, then, is infinite total potential information at high entropy while the dense nuclear fluid, the precursor to the universe, which formed out of Infinity, is stored information with low entropy.

In other words,

- The essence of Infinity is infinite potential information and high entropy.
- Information is the "ether" of Infinity.
- At some infinite time in Infinity, a self-organized "event" caused a coalescence of finite potential information (at high entropy and high disorder) into an extremely small "clot" of mass of finite store information (at low entropy and low disorder), at extremely high pressure and temperature. This was the formation of the precursor, dense, nuclear fluid.
- The precursor, dense, nuclear fluid was rapidly transformed into the pre-nascent gaseous universe at the same high pressure and temperature, of very small diameter, of low entropy, and high stored information.

- It was this finite prenascent gaseous universe, at extremely high temperature and pressure, and extremely small diameter, which then very rapidly expanded against the "vacuum" of Infinity, giving rise to the big bang. Thus was born the history of the universe.
- The universe is expanding, in gaseous form, much like, in a thermodynamics sense, an ideal gas expands.
- The expanding, gaseous universe is increasing in entropy content, in the same way an ideal gas increases in entropy as its volume increases and its pressure decreases. The universe, originally mostly stored information, becomes less stored information but more potential information as it expands and its internal structure becomes more random.
- The entropy laws as we know them on earth, if they are applicable to the expanding universe (seen as an ideal gas) would predict that the change in entropy content of the universe can be calculated from this formula.

$$\Delta S = 2.30 \, R \log_{10} \frac{V_2}{V_1} + 2.30 \, C_v \log_{10} \frac{T_2}{T_1}$$

where

ΔS = Change in entropy content for the universe, in going from state 1 to state 2

R = Ideal gas constant

V_1, V_2 = Volume of the universe in states 1 and 2

T_1, T_2 = Temperature of the universe in states 1 and 2

- Entropy, as informational entropy, can be determined for the universe from Shannon's informational entropy formula.

$$S = \sum p_i \log_{10} p_i$$

where

S = Informational Entropy

p_i = The condition and information content of the various internal parts of the universe

- These entropy equations can be used to track the increase in entropy content of the universe, from its birth as a precursor, dense nuclear fluid, through the big bang, to the present, and to its death at maximum entropy, as total potential information, and at the pressure and temperature of Infinity.

- When the universe achieves its maximum entropy and total potential information (its death), the curtain will part and the universe will return to the infinite potential information of Infinity. The universe will dissolve into Infinity and disappear. The stored information of its birth will have become the potential information of death, which is then merged with the infinite potential information of Infinity.
- There may be an Infinity of universes growing as points of stored information and low entropy. These are the singularities and discontinuities of Infinity.

Simple Models to Help Explain the Birth and Death of the Universe

(a) *The Birth of a Cloud Out of the Sky, and its Return to the Sky*

Within the sky there could, under certain circumstances, be formed a collection of vapor water molecules at the temperature of the sky. In sufficient numbers, these vapor water molecules can become liquid water molecules and form a cloud in the sky. The molecules have gone from vapor (high entropy) to liquid (low entropy) with a dramatic decrease in volume. This cloud of liquid water molecules is like a universe in the seemingly infinite sky. This cloud universe then spends its existence returning to its origins and is resorbed into the sky and disappears.

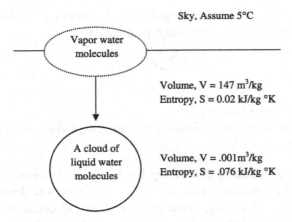

Sky, Assume 5°C

Vapor water molecules

Volume, V = 147 m³/kg
Entropy, S = 0.02 kJ/kg °K

A cloud of liquid water molecules

Volume, V = .001m³/kg
Entropy, S = .076 kJ/kg °K

Liquid water droplets form a cloud (like a universe in the sky)

(b) *Fireworks*

From within the sky-earth, under certain conditions, there could be formed a collection of unstable material (fireworks). Under another set of circumstances, these fireworks are ignited and eventually explode and spray burning particles in the sky. These rapidly burning particles become rapidly expanding gases, like universes in the seemingly infinite sky-earth. These particles-gases spend their existence returning to their origins, to be resorbed back into the sky-earth.

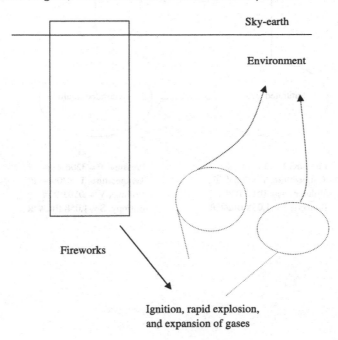

Sky-earth

Environment

Fireworks

Ignition, rapid explosion,
and expansion of gases

(c) *Saturated Steam-Saturated Water at its Critical Temperature and Pressure*

Under certain conditions of high pressure and high temperature, saturated liquid water can spontaneously transform itself into saturated water vapor, without outside intervention, and without any changes in its pressure, temperature, volume, or entropy. This saturated water vapor could then expand very rapidly against its environment, spending its existence returning to its origins and being resorbed and disappearing into its ambience.

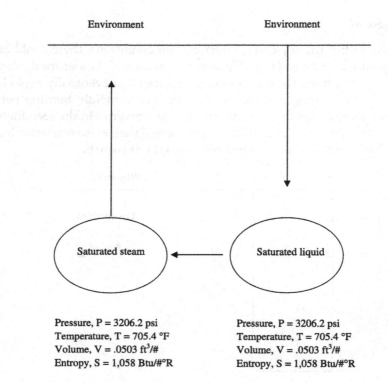

Environment   Environment

Saturated steam   ←   Saturated liquid

Pressure, P = 3206.2 psi   Pressure, P = 3206.2 psi
Temperature, T = 705.4 °F   Temperature, T = 705.4 °F
Volume, V = .0503 ft³/#   Volume, V = .0503 ft³/#
Entropy, S = 1,058 Btu/#°R   Entropy, S = 1,058 Btu/#°R

CHAPTER 18

Pressure, Volume, Temperature, Entropy Calculations: The Beginning, The Middle, and The End of the Universe

Physical properties of the universe

(a) The density of the universe = 3×10^{-31} gm/cm^3.

(b) Temperature of the universe is inversely proportional to its radius, or $T \alpha 1/R$, where R is the radius of the universe as a sphere, and T is the temperature of the universe.

(c) The present radius of the universe, R, is 20 billion light years, where it is assumed the universe is a sphere.

(d) The number of galaxies in the universe is 100 billion.

(e) The dimensions of a galaxy, as a short cylinder are shown below.

(f) The mass of a galaxy is 2×10^{33} gm.

(g) The age of the universe, A, is 10 billion years.

(h) The mass of the universe, m, is equal to the sum of the mass of all the galaxies, or m = (100 billion galaxies)$(2 \times 10^{33}$ gm$)$ = (1×10^{11}) (2×10^{33}) = 2×10^{44} gm.

(i) The density of a nuclear fluid is 10^{14} gm/cm^3.

(j) Age and temperature of the universe after its birth (see Table on next page).

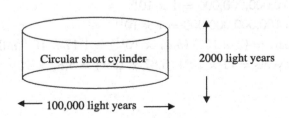

Circular short cylinder — 2000 light years

←— 100,000 light years —→

Age	Temperature
1 sec	10 billion °C
100 sec	1 billion °C
2 min	1 billion °C
1 hr	250 million °C
200,000 yr	6,000°C
300,000 yr	3,000°C
250 million yr	−100°C
10 billion yr	−270.27°C (2.73°K)

Numerical Conversion Factors

cm³ to miles³

1 mile = 5,280 ft × 12 in/ft × 2.54 cm/in = 1.61 × 10^5 cm

1 mile^3 = (1.61 × 10^5 cm)³ = 4.17 × 10^{15} cm^3

sec to yr

1 yr = 365 days × 24 hr/day × 60 min/hr × 60 sec/min
      = 3.15 × 10^7 sec

light year to mile

1 light year = 186,000 miles/sec × 3.15 × 10^7 sec/yr
             = 5.87 × 10^{12} miles

large numbers

One thousand = 1,000 = 1 × 10^3

One hundred thousand = 100,000 = 1 × 10^5

One million = 1,000,000 = 1 × 10^6

250 million = 250,000,000 = 2.5 × 10^8

One billion = 1,000,000,000 = 1 × 10^9

10 billion = 10,000,000,000 = 1 × 10^{10}

100 billion = 100,000,000,000 = 1 × 10^{11}

2,000 light years = (2 × 10^3) (5.87 × 10^{12}) = 1.17 × 10^{16} miles

100,000 light years = (1 × 10^5) (5.87 × 10^{12}) = 5.87 × 10^{17} miles

The Universe as an Ideal Gas

If the universe is considered as an ideal gas, then the ideal gas law for the universe can be written as

$$PV = mRT$$

where
P = pressure
V = volume
m = mass
R = ideal gas constant
T = temperature

If the ideal gas behaves as helium, then the ideal gas constant, R, is $2077 \, \text{Pa m}^3/\text{kg} \, °\text{K}$

where
Pa = pressure
m = meters
kg = kilograms
°K = degrees Kelvin

The gas constant may be adjusted as follows:

$$1 \, \text{Pa} = 1.45 \times 10^{-4} \text{ pounds per square inch (psi)}$$
$$1 \, \text{m}^3 = 35.3 \, \text{ft}^3 \times [1 \text{ mile}/5280 \, \text{ft}]^3 = 2.40 \times 10^{-10} \text{ miles}^3$$
$$1 \, \text{kg} = 1000 \, \text{gm} = 10^3 \, \text{gm}$$

Now the ideal gas constant, R, may be rewritten as

$$R = 2007 \, \text{Pa m}^3/\text{kg} \, °\text{K}$$
$$= 2.007 \times 10^3$$
$$\times \, [(1.45 \times 10^{-4} \, \text{psi})(2.4 \times 10^{-10} \text{ miles}^3)/(10^3 \, \text{gm})(°\text{K})]$$

or

$$R = 7.23 \times 10^{-14} \, \text{psi miles}^3/\text{gm} \, °\text{K}$$

Temperature–Volume Relationship of the Universe

It was given that temperature is inversely proportional to radius, or

$$T \alpha \, 1/R$$

where

α indicates proportionality and
T = temperature of the universe
R = radius of the universe as a sphere

The volume of a sphere is $V = 4/3 \, \pi R^3$, which combined with $T \alpha \, 1/R$ becomes

$$T \alpha \, 1/(3V/4\pi)^{1/3}$$

or

$$TV^{1/3} \alpha \, 1/(3/4\pi)^{1/3}$$

If the proportionality factor is K, then

$$TV^{1/3} = K \times 1/(3/4\pi)^{1/3}$$

or

$$TV^{1/3} = K'$$

where $K/(3/4\pi)^{1/3} = K'$. From T, V data for the universe, K' can be found.

Temperature of the Universe versus the Age of the Universe

The temperature (T) and age (A) data are available. Since these data span many orders of magnitude, it is best to express this relationship in log–log terms, using the base 10 for the logarithms. With the log data, we note it is possible the temperature and age variables will go to zero. The logarithm of zero is not defined so it is best to avoid this situation by, for example, arbitrarily modifying the temperature and age variable as follows.

$$\text{Modified Temperature, } T' = T(°C) + 300$$
$$\text{Modified age, } A' = A \text{ (yrs)} + (1 \times 10^{-8})$$

Birth of the Universe at A = 0 yrs

$$\text{At birth, with } A = 0, A' = 1 \times 10^{-8}$$
$$\text{and } \log_{10} A' = -8.00$$

Death of the Universe when the Temperature (T) Descends to Absolute Zero (0°K, or −273°C)

If it is assumed that the universe will die at $T = -273°C$, then

$$T' = -273 + 300 = 27$$

and

$$\log_{10} T' = 1.43$$

The Present Temperature (T) of the Universe

The present temperature of the universe is $T = -270.27°C$. This gives

$$T' = 29.7$$

and

$$\log_{10} T' = 1.47$$

A graph of $\log_{10} T'$ versus $\log_{10} A'$ for the Universe Yields the Present Age (A), the Death Age, and the Temperature (T) at Birth

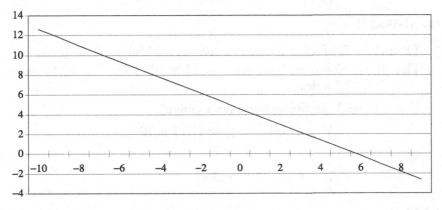

The present age of the universe from the regression line equation through all the data can be found at $\log_{10} T' = 1.47$. This yields a present age of 9.27 billion years.

The death age of the universe will occur at $\log_{10} T' = 1.43$ which yields the death age to be 11.18 billion years.

The temperature of the universe at birth is found from $\log_{10} A' = -8.00$, and is 19.42 billion °C.

Present Volume (V) of the Universe Calculated by Three Methods

(a) Method I

The density of the universe $= 3 \times 10^{-31}$ gm/cm^3

$$= (3 \times 10^{-31} \text{ gm/cm}^3)$$
$$(4.17 \times 10^{15} \text{ cm}^3/\text{miles}^3)$$
$$= 1.25 \times 10^{-15} \text{ gm/miles}^3$$

From the definition of density,

density of the universe

$= $ mass of the universe/volume of the universe

or

1.25×10^{-15} gm/miles^3

$= 2 \times 10^{44}$ gm mass of the universe/volume of the universe

or

$$V = \text{volume of the universe}$$
$$= 1.60 \times 10^{59} \text{ miles}^3$$

(b) Method II

The radius, R, of the universe as a sphere, is 20 billion light years.
Thus $R = 20$ billion light years $= (2 \times 10^{10}) \, (5.87 \times 10^{12})$ miles
$= 1.17 \times 10^{23}$ miles

The volume, V, of the universe (as a sphere)

$$V = 4/3\pi R^3 = 4/3\pi \, (1.17 \times 10^{23})^3$$

or

$$V = 6.70 \times 10^{69} \text{ miles}^3$$

(c) Method III

Consider the volume, V, of the universe to be completely filled by the 100 billion galaxies. The volume of a galaxy as a circular cylinder is

$$\text{Volume of a galaxy} = (\pi/4 \times D^2) \, (L)$$

where
$D = $ diameter of a galaxy $= 100,000$ light years
$L = $ thickness of the galaxy $= 2,000$ lights years

This yields

$$D = (1 \times 10^5)\,(5.87 \times 10^{12}) = 5.87 \times 10^{17} \text{ miles}$$
$$L = (2 \times 10^3)\,(5.87 \times 10^{12}) = 1.17 \times 10^{16} \text{ miles}$$

and the volume of a galaxy $= (\pi/4)(5.87 \times 10^{17})^2(1.17 \times 10^{16})$
$= 3.16 \times 10^{51} \text{ miles}^3$

The volume of the universe, V, completely filled with 100 billion galaxies is

$$V = (100 \text{ billion galaxies}) \, (3.16 \times 10^{51}) \text{ miles}^3$$
$$V = (1 \times 10^{11})\,(3.16 \times 10^{51}) = 3.16 \times 10^{62} \text{ miles}^3$$

Evaluation of K' from $TV^{1/3} = K'$

At the present time, the temperature, T, of the universe is $-270.27°C$ (2.73°K). Choose as the present volume, V, of the universe the largest number obtained here, $V = 6.70 \times 10^{69} \text{ miles}^3$.

From $TV^{1/3} = K'$, in the units of T, °K, and V, miles³, K' can be calculated.

$$2.73(6.70 \times 10^{69})^{1/3} = K'$$

or

$$K' = 5.13 \times 10^{23}$$

with $TV^{1/3} = 5.13 \times 10^{23}$ and assuming K' is constant over the lifetime of the universe, then the volume of the universe can be calculated at its various stage of life: from birth, to maturity, to the present. All that is needed is the temperature data at given ages of the universe.

The Volume, V, of the Universe, From Birth to Death

From $TV^{1/3} = 5.13 \times 10^{23}$ where T is °K and V is miles³. The temperature, T, data can be converted to °K by the definition.

$$T, °K = T°C + 273$$

or

$$T, °K = T°C + .273 \times 10^3$$

The following table shows the temperature, T, data and the calculated volume, V, for the universe.

Age	Years	Temperature, T, °C	°K	Volume, $V^{1/3}$	V, miles^3
Birth	0	19.42 billion (1.94×10^{10})	1.94×10^{10}	2.64×10^{13}	1.84×10^{40}
1 sec	3.17×10^{-8}	10 billion (1×10^{10})	1×10^{10}	5.13×10^{13}	1.135×10^{41}
100 sec	3.17×10^{-6}	1 billion (1×10^{9})	1×10^{9}	5.13×10^{14}	1.35×10^{44}
2 min	3.81×10^{-6}	1 billion (1×10^{9})	1×10^{9}	5.13×10^{14}	1.35×10^{44}
1 hr	1.14×10^{-4}	250 million (2.5×10^{8})	2.5×10^{8}	2.05×10^{15}	8.60×10^{45}
200,000 yr	2.0×10^{5}	6,000 (6×10^{3})	6.27×10^{3}	8.18×10^{19}	5.47×10^{59}
300,000 yr	3.0×10^{5}	3,000 (3×10^{3})	3.27×10^{3}	1.57×10^{20}	3.87×10^{60}
250 million yr	2.5×10^{8}	−100 (-1×10^{2})	1.73×10^{2}	2.96×10^{21}	2.59×10^{64}
9.27 billion yr (present)	9.27×10^{9}	−270.27 (-2.70×10^{2})	2.73×10^{0}	—	6.7×10^{69}
11.18 billion yr (death)	1.11×10^{10}	−273 (-2.73×10^{2})	0	—	1.0×10^{70}*

*obtained from a linear extrapolation of the last two Age–Volume data points (2.5×10^{8}, 2.59×10^{64} and 9.27×10^{9}, 6.70×10^{69}).

The Pressure of the Universe from Birth to Death

The universe, as an ideal gas, would be governed by the ideal gas law,

$$PV = mRT$$

where
m = mass of the universe = 2×10^{44} gm (assumed constant)
R = ideal gas law constant = 7.23×10^{-14} psi miles^3/gm°K and
T = temperature, °K
V = volume, miles^3
P = pressure, psi

For each age of the universe, the temperature and volume are known. In the ideal gas equation, $PV = mRT$, only P is unknown and can be calculated for each age of the universe. The table shown here summarizes these calculations.

Age	T, °K	V, miles^3	P, psi
Birth	1.94×10^{10}	1.84×10^{40}	15.2
1 sec	1×10^{10}	1.35×10^{41}	1.07
100 sec	1×10^{9}	1.35×10^{44}	1.07×10^{-4}
2 min	1×10^{9}	1.35×10^{44}	1.07×10^{-4}
1 hr	2.5×10^{8}	8.60×10^{45}	4.20×10^{-7}
200,000 yr	6.27×10^{3}	5.47×10^{59}	1.66×10^{-25}
300,000 yr	3.27×10^{3}	3.87×10^{60}	1.22×10^{-26}
250 million yr	1.73×10^{2}	2.59×10^{64}	9.66×10^{-32}
9.27 billion yr (present)	2.73×10^{0}	6.7×10^{69}	5.89×10^{-39}
11.18 billion yr (death)	0	1.0×10^{70}	0

Notes

Birth temperature was obtained previously from a graph of Temperature versus Age, when the age data are extrapolated to where Age is zero.

The present age of the universe is obtained from the same graph. We found the age where the temperature of the universe is its present temperature, 2.73°K.

The projected death age of the universe was obtained from the same graph. We found the age where the temperature of the universe reaches 0°K.

The pressure and volume of the universe are obtained from the ideal gas law, and the temperature versus age data available in the literature, and the published observations that the temperature of the universe is inversely proportional to its radius (if the universe were a sphere).

The Formation of the Prenascent Gaseous Universe From the Precursor Dense Nuclear Fluid

The density of the precursor dense nuclear fluid is 10^{14} gm/cm^3. Density is mass/volume, or

Density of the precursor dense nuclear fluid

= Mass of the precursor dense nuclear fluid/

Volume of the precursor dense nuclear fluid

Assume the mass of the precursor dense nuclear fluid is equal to the subsequent mass of the universe. This is, 2×10^{44} gm. Then Density of the precursor dense nuclear fluid $= 2 \times 10^{44}$ gm/Volume of the precursor dense nuclear fluid and therefore,

Volume of the precursor dense nuclear fluid
$$= 2 \times 10^{44}/1 \times 10^{14}$$
$$= 2 \times 10^{30} \, cm^3$$
$$= (2 \times 10^{30} \, cm^3) \, (1 \, miles^3/4.17 \times 10^{15} \, cm^3)$$
$$= 4.8 \times 10^{14} \, mile^3$$

It is the mass and volume of the precursor dense nuclear fluid, 2×10^{44} gm and 4.8×10^{14} mile^3 which becomes the prenascent gaseous universe with mass and volume the same as the precursor dense nuclear fluid.

The Phase Transformation, from the Precursor Dense Nuclear Fluid to the Prenascent Gaseous Universe

The mechanism for this phase transformation may be similar to the phase change of saturated liquid water to saturated gaseous steam at its critical temperature and pressure: the critical temperature is 705.4°F and the critical pressure is 3206.2 psi. Under these conditions, the volumes of saturated liquid water and saturated gaseous steam are the same. The enthalpy (energy content) of the liquid water and gaseous steam are the same. In other words, at this critical temperature and pressure, the saturated liquid water can transform itself into saturated gaseous steam without outside intervention. The phase change can occur in either direction, liquid to gas, or gas to liquid, without the need for additional heat transfer. One phase can become the other without any change in pressure, volume, or temperature. The entropy content of the saturated liquid water and the saturated gaseous steam are also the same.

This discussion of the properties of water and steam at a critical temperature and pressure offers a possible mechanism for explaining how the precursor dense nuclear fluid might become the prenascent gaseous universe. Suppose the precursor dense nuclear fluid emerges from infinity, at its critical temperature and pressure (like the water-steam example). This means the precursor dense nuclear fluid can execute a phase change without outside intervention. Thus the precursor dense nuclear fluid becomes the prenascent gaseous universe with properties the same as the precursor dense nuclear fluid. This is accomplished without the need for energy exchange or heat transfer from beyond its boundaries. Now the prenascent gaseous universe, at high temperature and pressure, and the low entropy content of stored information, begins its very rapid expansion (the big bang?) against the non-resistive ambient Infinity of low temperature and pressure, and the high entropy content of infinite potential information. With time, the gaseous universe expands and cools, decreasing in pressure and temperature, and decreasing in stored information, increasing in potential information and increasing in entropy. This will continue until the gaseous universe's temperature, pressure, and entropy content return to those values of the ambient Infinity. Then the gaseous universe will return to (or dissolve back into) Infinity and "disappear". The universe will have become potential information again and blended into Infinity, much as a cloud in the sky will eventually disappear or be resorbed into the sky.

Pressure, Volume, and the Temperature of the Prenascent Gaseous Universe

The volume of the prenascent gaseous universe is the same as the volume of the precursor dense nuclear fluid, and is equal to 4.8×10^{14} miles^3.

$$TV^{1/3} = 5.13 \times 10^{23}$$

where

T = temperature of the prenascent gaseous universe, °K

or

$$T = 5.13 \times 10^{23}/V^{1/3} = 5.13 \times 10^{23}/(4.8 \times 10^{14})^{1/3}$$
$$= 6.55 \times 10^{18}°K \cong 6.55 \times 10^{18}°C$$

Since the prenascent gaseous universe is also assumed to behave as an ideal gas, then

$$PV = mRT$$

or

$$P\,(4.8 \times 10^{14} \text{ miles}^3) = (2 \times 10^{44} \text{ gm})\,(7.23 \times 10^{-14})$$
$$\times (6.55 \times 10^{18}°K)$$

or

P = pressure of the prenascent gaseous universe = 1.97×10^{35} psi

Pressure, Volume, and Temperature of the Precursor Dense Nuclear Fluid which Formed Out of Infinity

Since the precursor dense nuclear fluid and the prenascent gaseous universe are at their critical temperature and pressure, then the temperature of the precursor dense nuclear fluid and the prenascent gaseous universe will be the same. The pressure will be the same for both phases as will the volume. This gives for the precursor dense nuclear fluid,

$$\text{Volume, } V = 4.80 \times 10^{14} \text{ miles}^3$$
$$\text{Pressure, } P = 1.97 \times 10^{35} \text{ psi}$$
$$\text{Temperature, } T = 6.55 \times 10^{18}\text{°C}$$

The entropy content of the precursor dense nuclear fluid and the prenascent gaseous universe will have the same low value.

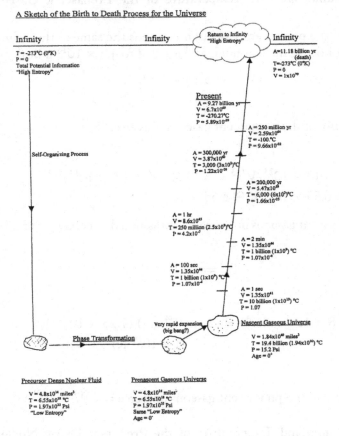

OUR AGING, EVOLVING UNIVERSE

A Sketch of the Birth to Death Process for the Universe

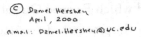

© Daniel Hershey
April, 2000
email: Daniel.Hershey@uc.edu

Entropy of the Universe

Using the example of the letters in the book, organized into words and lines and pages, arranged in order and bound, we conclude this is stored information, at low informational entropy.

If we then cut the pages into lines, the lines into words, the words into individual letters, and pile them on the floor, we conclude this is potential information, at high informational entropy.

Informational entropy, in these two situations, would seem to depend on the density of the letters, i.e., the number of letters per unit volume occupied by the letters. The book has a high density of letters and hence is stored information with low entropy. The pile of letters scattered on the floor has a lesser density of letters.

It is logical, therefore, to say that the informational entropy is inversely proportional to the informational density, or

$$S = \text{Informational Entropy} = K/D$$

where

K = a proportionality constant
D = informational density

The universe, in the vicinity of death, is at maximum informational entropy, is in total disorder, in total randomness. This total dispersion means everything within the universe is disassociated and unattached and unbound.

In Shannon's formula for informational entropy, it can be shown that maximum entropy, S_{max}, for this final state of the universe will be

$$S_{max} = \log_{10} m$$

where

$$m = \text{the mass of the universe} = 2 \times 10^{44}\, \text{gm}$$

The informational density, D, of the universe in the vicinity of death is $D = m/V$ = mass of the universe, gm/volume of the universe, miles^3 = $2 \times 10^{44}/1 \times 10^{70} = 2 \times 10^{-26}$ gm/miles^3.

In the vicinity of death, and from $S = K/D$ and Shannon's formula for maximum entropy, we obtain $K = D S_{max}$ where

$$S_{max} = \log_{10}(2 \times 10^{44})$$

or

$$K = (2 \times 10^{-26})[\log_{10}(2 \times 10^{44})]$$

where K is assumed constant over the lifetime of the universe.

For the precursor dense nuclear fluid,

$$V = 4.8 \times 10^{14} \text{ miles}^3$$
$$m = 2 \times 10^{44} \text{ gm}$$
$$D = m/V = 2 \times 10^{44}/4.8 \times 10^{14} = 4 \times 10^{29} \text{ gm/miles}^3$$

and

$$S = K/D = (2 \times 10^{-26})[\log_{10} 2 \times 10^{44}]/4 \times 10^{29}$$

or

$$S \cong 0$$

This calculation shows that the universe (as the precursor dense nuclear fluid which emerged from Infinity by a self-organizing process) has essentially zero informational entropy and is all stored information. It then proceeds through the various stages of the universe lifecycle until in the vicinity of death, the informational entropy level of the universe is at a maximum value, $S_{max} = \log_{10} 2 \times 10^{44}$. This informational entropy is in the form of complete potential information. In this moribund state, the universe with its pressure and temperature equal to that of Infinity, meets the necessary conditions for its return to Infinity. Thus the curtain parts and the universe disappears (is resorbed) into Infinity. This potential information, "borrowed" from Infinity by the formation of the precursor dense nuclear fluid, has been returned.

From Shannon's informational entropy formula, we estimate the ambient entropy of Infinity (in Shannon units) to be

$$S = \log_{10} m,$$

or

$$S = \log_{10} 2 \times 10^{44} = 44.301$$

Classic Thermodynamics Calculations

The ideal gas law constant, R, can be given as

$$R = 7.23 \times 10^{-14} \text{ psi miles}^3/\text{gm}°\text{K}$$

if it is assumed that the universe is essentially helium. If the thermal conductivity for the universe is also that of helium, then k = thermal conductivity of the universe = 1.67, in units similar to those given for R.

The ideal gas thermodynamics relationships are assumed to hold, so that

$$K = 1.67 = C_p/C_v$$

where C_p and C_v are the heat capacities of the universe(helium) at constant pressure and volume respectively. We also know for an ideal gas

$$C_p - C_v = R$$

or

$$C_p/C_v - 1 = R/C_v$$

or

$$1.67 - 1 = R/C_v$$

or

$$0.67 = R/C_v$$

and rearranging, $C_v = R/0.67 = 1.49R = 1.49(7.23 \times 10^{-14})$

For an ideal gas $= 1.08 \times 10^{-13}$ psi miles^3/gm°K

$$\Delta S = R \ln V_2/V_1 + C_v \ln T_2/T_1$$

Converting ln(natural logarithms) into $\log_{10}$(logarithms to the base 10) we have

$$\log_{10} N = 0.434 \ln N$$

or

$$\ln N = 2.30 \log_{10} N$$

and

$$\Delta S = R(2.30 \log_{10} V_2/V_1) + C_v(2.30 \log_{10} T_2/T_1)$$

or

$$\Delta S = 2.30R \log_{10} V_2/V_1 + 2.30 C_v \log_{10} T_2/T_1$$

or, with

$$R = 7.23 \times 10^{-14},$$

we get

$$2.30R = 1.66 \times 10^{-13}$$

and

$$2.30C_v = 2.48 \times 10^{-13}$$

or

$$\Delta S = (1.66 \times 10^{-13}) \log_{10} V_2/V_1 + (2.48 \times 10^{-13}) \log_{10} T_2/T_1$$

Calculating the entropy change for the universe as an ideal gas, in going from the prenascent gaseous universe to death, we get:

(1) From $t = 0-$ to $t = 0+$
Prenascent gaseous   Nascent Gaseous
Universe   Universe
(1)   (2)

$$\Delta S = (1.66 \times 10^{-13}) \log_{10} 1.84 \times 10^{40}/4.8 \times 10^{14}$$
$$+ (2.48 \times 10^{-13}) \log_{10} 1.94 \times 10^{10}/6.55 \times 10^{18}$$
$$= 2.14 \times 10^{-12}$$

(2) From $t = 0+$ to $t = 9.27$ billion years
Nascent Gaseous (9.27×10^9)
Universe   Present

$$\Delta S = (1.66 \times 10^{-13}) \log_{10} 6.7 \times 10^{69}/1.84 \times 10^{40}$$
$$+ (2.48 \times 10^{-13}) \log_{10} 2.73 \times 10^0/1.94 \times 10^{10}$$
$$= 2.47 \times 10^{-12}$$

(3) From $t = 9.27$ billion years   to $t = 11.18$ billion years
(9.27×10^9) (11.18×10^9)
Present   Death

To get ΔS from the present to the death, we use a linear extrapolation from the Nascent Gaseous Universe to the Present to get ΔS from present to death.

$$\Delta S = \frac{\text{time, death}}{\text{time, Present}} \times \Delta S(t = 0^+ \text{ to Present})$$

or

$$\Delta S = [(11.18 \times 10^9 - 9.27 \times 10^9)/9.27 \times 10^9] \times 2.47 \times 10^{-12}$$

and

$$\Delta S = 0.51 \times 10^{-12}$$

(4) From $t = 0-$ to $t = 11.18$ billion years
Prenascent Gaseous (11.18×10^9)
Universe   Death

$\Delta S = \Delta S$ (Prenascent to Nascent) $+ \Delta S$ (Nascent to Present)
       $+ \Delta S$ (Present to Death)
     $= 2.14 \times 10^{-12} + 2.47 \times 10^{-12} + 0.51 \times 10^{-12}$

or

$$\Delta S = 5.12 \times 10^{-12}$$

It was shown previously that the entropy content of the precursor dense nuclear fluid is essentially zero, and therefore, the precursor gaseous universe is also essentially zero. This means the universe, in evolving from the precursor dense nuclear fluid to death, generates an entropy production of $\Delta S = 5.12 \times 10^{-12}$ (in classical thermodynamic units associated with R, C_p, T, V, P). This is the maximum entropy of the universe, before it returns to ambient Infinity, and is also the ambient entropy level of Infinity.

Thus we have two numbers associated with the entropy level in Infinity. They express the same concept, but are in different units.

Entropy Level of Infinity $= 44.301$ (Shannon's units)

or

Entropy Level of Infinity
     $= 5.12 \times 10^{-12}$ (Classical Ideal Gas Thermodynamics Units)

Coda: If All This Is True, Then ...

Now we can begin to understand the significance of our human existence on earth, and explore the questions of how all of this relates to the concept of God. We are an integral part of an increasingly complex society — of earth, of our galaxy, of the universe, and of Infinity, which is beyond everything. Understanding the hierarchy of things places our lives and our accomplishments in perspective. Understanding our origins and our limitations produces a certain humility which can lead to an appreciation of our very finite stay on earth. Understanding all of this can perhaps encourage a morality which can yield a permanent peace on earth.

I know I am but an infinitesimal cog in a larger, infinite order of things. We shall be born, live and die, and in being, alter the universe. The march towards Infinity will be affected ever so slightly by my record of existence. Very few will be aware of my history. But I know, and by knowing, I endeavor to construct a meaningful record.

We believe in God because there is comfort in believing that God fills the gaps in our universe. God clearly represents our understanding of our limitations. But we need to reconcile our concept of God with the newer idea of Infinity. When I say "God", you are not allowed to "see" God, you are forbidden to sense anything tangible. For if you do, and then I can ask the next question, "What is beyond God?" And if you respond that you see a bigger, more powerful God, then I can ask "What is beyond that?" And where does it end? In Infinity, of course. So God and Infinity are linked.

Now if I say "Universe" and again you see something tangible, with dimensions and a boundary, then I once again can ask, "What is beyond the universe?" And so we are led to the possibility of the multiple universes, like multiple clouds in the sky. As many as there are, these universes may coexist, or may coalesce, and eventually return to Infinity separately at different times or together at one time.

If all this is true, then wars are senseless. Whatever the gains of the mass killing, it cannot be justified in the long run. We are only temporary inhabitants of a temporary earth, of a temporary galaxy, of a temporary universe. Why not work for the betterment of our families? Why not fill our days with meaningful work? Why not love ourselves and each other.

If there is anything which ought to remain constant, it is the sanctity of life. My life belongs to me; it is mine to use as I wish. No President of any country has the right to demand my life for a cause. I don't relinquish my claim, that even under the gravest of circumstances; no President can force me to give up my life. My life is independent of local politics; my life is global. My life is universal.

Our universe may end in two billion years. Don't make any long range plans.

May your entropy be small.

Author Biography

Professor Daniel Hershey is an Emeritus Professor at the University of Cincinnati and a Fellow of the Graduate School, with a B.S. from the Cooper Union in New York City and a Ph.D. from the University of Tennessee.

During the 1970s, Prof. Hershey served as Assistant to the then President of the University of Cincinnati, Warren Bennis.

Prof. Hershey received two Fulbright Fellowships, two Tau Beta Pi teaching awards, a clinical research award from the American Society of Bariatric Physicians, NIH grants for blood flow and blood oxygenation studies, and a first-place award from the Cincinnati Editors Association for a short story entitled "A Letter to Michael". From 1970 to 1980, he presented two radio programs, *Must We Grow Old* and *Everyday Science*, on WGUC, a radio station in Cincinnati. He also has a patent for his whole body calorimeter which measures the basal metabolic rate of humans.

Prof. Hershey's research and teaching at the University of Cincinnati explores the aging process for humans, corporations and the universe, and he has developed an entropy theory to help explain the aging process. He is the author of 14 books, which include *Hershey Corporate Lifecycle Assessment, Diagnosing a Corporate Bureaucracy, Lifespan Potential, Longevity, Obesity, and Fitness in Aging Humans, Must We Grow Old (From Pauling, to Prigogine, to Toynbee), The Universe and Beyond, Entropy, Infinity, and God,* and *Lifespan and Factors Affecting It.*

Index